SpringerBriefs in Plant Science

SpringerBriefs present concise summaries of cutting-edge research and practical applications across a wide spectrum of fields. Featuring compact volumes of 50 to 125 pages, the series covers a range of content from professional to academic. Typical topics might include:

- A timely report of state-of-the art analytical techniques
- A bridge between new research results, as published in journal articles, and a contextual literature review
- A snapshot of a hot or emerging topic
- An in-depth case study or clinical example
- A presentation of core concepts that students must understand in order to make independent contributions

SpringerBriefs in Plant Sciences showcase emerging theory, original research, review material and practical application in plant genetics and genomics, agronomy, forestry, plant breeding and biotechnology, botany, and related fields, from a global author community. Briefs are characterized by fast, global electronic dissemination, standard publishing contracts, standardized manuscript preparation and formatting guidelines, and expedited production schedules.

More information about this series at http://www.springer.com/series/10080

Girdhar K. Pandey • Sibaji K. Sanyal

Functional Dissection of Calcium Homeostasis and Transport Machinery in Plants

 Springer

Girdhar K. Pandey
Department of Plant Molecular Biology
Benito Juarez Road, Dhaula Kuan
University of Delhi South Campus
New Delhi, Delhi, India

Sibaji K. Sanyal
Department of Plant Molecular Biology
Benito Juarez Road, Dhaula Kuan
University of Delhi South Campus
New Delhi, Delhi, India

ISSN 2192-1229   ISSN 2192-1210   (electronic)
SpringerBriefs in Plant Science
ISBN 978-3-030-58501-3   ISBN 978-3-030-58502-0   (eBook)
https://doi.org/10.1007/978-3-030-58502-0

This Springer imprint is published by the registered company Springer Nature Switzerland AG
The registered company address is: Gewerbestrasse 11, 6330 Cham, Switzerland

This work is dedicated to
Sir Michael John Berridge,
for opening the "gates" of calcium signaling

Preface

Plants are sessile, and this is a major limitation in terms of motility in response to favorable and unfavorable growth conditions, despite the fact that these are dominant species on Earth. To circumvent this problem, plants use a signaling system that can cater to a wide array of stimuli. These signaling systems particularly help plants to respond when challenged with stress (abiotic or biotic). With the advent of holistic approaches to understand the signal transduction mechanisms, many researchers have significantly contributed to the field of cellular communication. Their efforts have allowed us to draw a bigger picture of signaling components and pathways. One of the prime important molecules involved in signal transduction is calcium. Anthony Trewavas had once named one of his articles "Le Calcium, C'est la Vie: Calcium Makes Waves" which can be translated to "Calcium (Ca^{2+}) is life." In the article, he discussed the significance of Ca^{2+} waves in the cell and the role of Ca^{2+} in plant intelligence. True to this, Ca^{2+} in the living organism can play a dual role – as a structural and as a signaling component. The role of Ca^{2+} signaling becomes more important in "sessile" plants (compared to animals). Further, it would be wrong to imagine Ca^{2+} signaling playing a role only to mitigate stress; it is also an essential factor in normal growth and development. The evolution of Ca^{2+} signaling is a fascinating story. The extrusion of cytotoxic Ca^{2+} from cytosol has made it an effective signaling molecule. This particular event (extrusion of Ca^{2+} from cytosol) led to the development of the phenomenon of Ca^{2+} homeostasis, and with it, proteins evolved to maintain this homeostasis.

Initial studies identified Ca^{2+} only as a structural component of the cell. Later, animal biologists found that Ca^{2+} could work as a signaling molecule. Following this discovery, plant biologists also confirmed the same phenomenon occurring in plants. Then, with the help of better experimental approaches (which have refined over a period of time), valuable information has been gathered, which has enhanced our understanding of Ca^{2+} signaling and homeostasis. Over the years, more plant biologists have worked in this field and also done a commendable job to look at plant signaling with an integrative approach. We now know that plants respond to stimuli (favorable or unfavorable) by translating them into effective cellular messages that involve a plethora of pathways. The messages are transduced at the gene level (by regulation of expression) or at a physiological level (by altering the plant physiology). The Ca^{2+} signaling pathway plays a pivotal role in a plant's response to stimuli. It is one of the key messengers to perceive stimulus and pass the information downstream in a signaling pathway to bring about a response.

As we move towards a challenging future, we need future-ready plants (crop plants in general) to survive and also for our endeavors in space research. Breeding and engineering crops with essential traits will be a necessity for the future. For this purpose, the basic understanding of stimulus perception becomes of utmost importance. As "Ca^{2+} signaling is a network that connects the plant cell," an updated knowledge of Ca^{2+} homeostasis events in general and the transport elements in particular is of utmost importance. So with this book, *Functional Dissection of Calcium Homeostasis and Transport Machinery in Plants*, we bring forward the current understanding of the field of Ca^{2+} signaling and the transport elements involved in the maintenance of Ca^{2+} homeostasis in the cell. We provide brief glimpses of all nuances of Ca^{2+} signaling here – from genesis of Ca^{2+} signaling to the players involved in maintaining homeostasis and also the tools used to identify and study them.

In **Chap. 1,** we discuss the dual role played by Ca^{2+}, as a nutrient and as a second messenger, and we give a brief outline of the evolution of this ion as a second messenger and the elements involved in the Ca^{2+} signaling pathway. Ca^{2+} concentration in the cytosol needs a stringent regulation. This is also essential for Ca^{2+} to play its role as a signaling molecule. So in **Chap. 2,** we discuss the paradigm of Ca^{2+} homeostasis and Ca^{2+} reserves inside the cell. We also give a very brief overview of the players involved in Ca^{2+} homeostasis. The Ca^{2+} homeostasis event manages the transient rise (and fall) of Ca^{2+} in the cytosol. This transient rise (Ca^{2+} signature) is a result of (any) stimuli and carries encoded messages. In **Chap. 3,** we discuss the Ca^{2+} signature hypothesis and the organellar Ca^{2+} signature. We also give a brief overview of the cross talk of Ca^{2+} with other pathways. The encoded messages in the Ca^{2+} signature need to be decoded. In **Chap. 4,** we discuss the plethora of Ca^{2+} binding proteins present in a plant cell that can decode and transduce the messages encoded in a Ca^{2+} signature. The ion channels mediate Ca^{2+} entry into the cell. Initial studies of the plant had identified several ion channels that are modulated by membrane voltages. In **Chap. 5,** we talk about these ion channels, which are known through their voltage-dependent (or independent) conductance, but the corresponding proteins are yet to be identified. We also discuss two-pore channel 1, a major vacuolar ion channel putatively involved in enhancing the Ca^{2+} signals. Besides membrane voltages, ligands (like amino acids and cyclic nucleotides) can also modulate ion channels for mediating Ca^{2+} release into the cytosol. In **Chaps. 6** and **7,** we talk about glutamate-like receptors and cyclic nucleotide-gated channels, respectively. Besides these, plant cells also have some very unique channels. The annexins and mechanosensitive channels are different from the other ion channels. We talk about them in **Chap. 8.** The Ca^{2+} brought inside the plant cell must be under tight regulation and quickly effluxed from the cytosol. This work is performed by a family of ATPases and Ca^{2+} antiporters discussed in **Chap. 9**. Finally, in **Chap. 10,** we elaborate on the major techniques used to study the Ca^{2+} transport elements. We also discuss two very unique roles of Ca^{2+} as a long-distance message carrier and its role in plant memory. We conclude the work here with some important questions for the near future.

The information provided in the write-up is an attempt to give the reader a very brief and quick glimpse of the field of Ca^{2+} signaling. The very brief nature of the work was challenging as we had to include several aspects of Ca^{2+} signaling and transporters. We had to limit ourselves, and as such many seminal works were not discussed at length. We have cited many excellent reviews and books (and book chapters) in the text if the reader decides to explore more on a certain topic. We hope this concise and crisp study will act as a guide for researchers and students who look to indulge themselves in understanding "calcium transport and signaling" at the holistic level.

New Delhi, Delhi, India   Girdhar K. Pandey
                                                                                              Sibaji K. Sanyal

Acknowledgment

We are thankful to the Department of Biotechnology (DBT), Science and Engineering Research Board (SERB), Council for Scientific and Industrial Research (CSIR), and Delhi University (R&D grant), India, for supporting the research work in the GKP lab. Financial support from the DBT-RA Program in Biotechnology and Life Sciences (India) is gratefully acknowledged by SKS.

Contents

Chapter 1
Calcium- from Nutrition to Signaling

Contents

Introduction

Plants require mineral nutrients for their optimal growth. These are supplied by macronutrients (viz. nitrogen, sulphur, phosphorus, magnesium, potassium and calcium). Plants also require micronutrients like iron, manganese, copper, zinc, nickel, molybdenum, boron and chlorine (Broadley et al. 2012; Hawkesford et al. 2012; Kirkby 2012). Among these important mineral elements, calcium (Ca^{2+}) stands out as it can perform a dual role -one as the already stated nutritional role, and the -other as a very important role of second messenger (Trewavas and Malho 1998; Sanders et al. 1999; Berridge et al. 2000; Dodd et al. 2010; Hawkesford et al. 2012). We look into these aspects of Ca^{2+} in plant biology in this chapter.

Ca²⁺ as a Nutrient

Ca^{2+} is one of the most abundant mineral elements present in the soil (very abundant in the earth and estimated 1650 μM in about 20 cm soil solution in cultivable land) (Maathuis 2009; Marschner and Rengel 2012). Plant uptake Ca^{2+} readily from the soil with neutral (even slightly basic) pH, and Ca^{2+} absorption is hindered in soil with an acidic pH (Hawkesford et al. 2012). Increase in soil heavy metal content can also hinder in the Ca^{2+} uptake (Hawkesford et al. 2012). Plants take Ca^{2+} in the form of ions through the ion channels located in the roots. These channels are usually

G. K. Pandey, S. K. Sanyal, *Functional Dissection of Calcium Homeostasis and Transport Machinery in Plants*, SpringerBriefs in Plant Science,
https://doi.org/10.1007/978-3-030-58502-0_1

nonselective (i.e., wide range of minerals, as well as Ca^{2+}, can be taken in) and are depolarization activated channels (DACC) or hyperpolarization-activated channels (HACC) (Maathuis 2009; Thor 2019). We discuss more of these channels in Chaps. 2 and 5. The majority of the Ca^{2+} taken-up by root is transported to shoot through the xylem system. As such the leaves are the primary stores of Ca^{2+} (about 70%) and the rest is present in roots (Conn and Gilliham 2010).

The transport of Ca^{2+} through the root to leaves is through the xylem (due to transpiration) and not through the phloem (Ca^{2+} is immobile in phloem) (González-Fontes et al. 2017). In the roots, Ca^{2+} may follow the apoplastic or symplastic movement to reach the xylem depending on the species (González-Fontes et al. 2017). The Ca^{2+} movement begins from the epidermis, crosses the cortex and stops at the casparian strips of the endodermis (Thor 2019). Here Ca^{2+} enters the cytosol of endodermis (via some channel proteins) and is exported to the xylem by Ca^{2+} ATPases or Ca^{2+}/H^+ antiporters (more on them later in the book). Once the xylem has transported it to leaves (through CNGC channels), it is stored in mesophyll cells, trichomes or epidermal cells (Thor 2019). Here again, storage preference is species-specific (González-Fontes et al. 2017). The preferred storehouse inside the cell is vacuole, which accumulates the highest concentration of Ca^{2+} (Stael et al. 2012). Figure 1.1 provides a hypothetical model of Ca^{2+} uptake in plants.

As we have already mentioned certain conditions may interfere with the plant Ca^{2+} uptake (acidic soil or presence of heavy metals that impede Ca^{2+} absorption from root), this can result in Ca^{2+} deficiency (Hawkesford et al. 2012; Thor 2019). Ca^{2+} deficiency is more commonly seen in young leaves and fruits (Thor 2019). Ca^{2+}, hence serves a wide variety of nutritional requirement in plants. Ca^{2+} is an important component of the cell wall and is bound to pectins to strengthen the cell wall. The presence of Ca^{2+} in the cell wall is also important for maintaining cell wall integrity and replacement of Ca^{2+} in the cell wall can either result in solute leakage or metal toxicity. Ca^{2+} also plays a significant role in cell elongation -best examples are elongation of the pollen tubes and root hairs. The mucilage secretion and callose formation also depend on the apoplastic Ca^{2+}. The Ca^{2+} stored in the vacuole is important for maintaining cellular ion-balance (by forming Ca-oxalate with the oxalate generated during nitrate reduction) (Hawkesford et al. 2012). External Ca^{2+} can protect plants against pathogens (Thor 2019). Although Ca^{2+} competes with magnesium (Mg^{2+}) for binding with enzymes and possibly interfering with enzyme activity, it can still activate enzymes, e.g., alpha-amylases and suggest that Ca^{2+} also play a role in activating enzyme (Hawkesford et al. 2012).

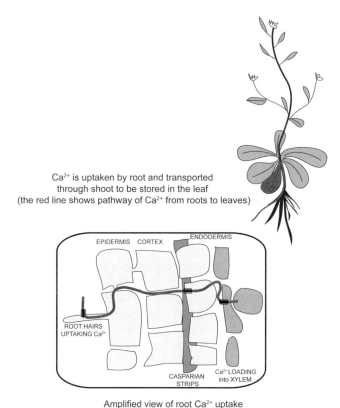

Fig. 1.1 **A hypothetical model of Ca²⁺ uptake by the plants.** Root plasma membrane-localized depolarization activated Ca²⁺ channel (DACC) and hyperpolarization-activated Ca²⁺ channel (HACC) uptake Ca²⁺ from the soil. This majorly moves through the epidermis, cortex and enter the cytoplasm of endodermis when it encounters the casparian strips. From the endodermal cytosol, Ca²⁺ is loaded into the xylem using Ca²⁺-ATPases. Through the xylem, Ca²⁺ travels to leaves where it is majorly stored. The magnified root structure shows Ca²⁺ travelling path with the red line. In the plant, the red line depicts the Ca²⁺ movement through the stem to leaves. One leaf is highlighted in red color to depict it as the major Ca²⁺ storage in plants. Ca²⁺ transport through root is inspired by (Thor 2019)

The Transition of Calcium from Nutrient to Messenger in the Cell

Decades of research have proved that this ion plays a very significant role in signal transduction events in the cell. It serves as an important signaling molecule transducing information across the cell by connecting the target and the information center. This particular role of Ca^{2+} has been appreciated significantly in recent times (Hepler and Wayne 1985; Trewavas 1999; Clapham 2007; Dodd et al. 2010; Kudla et al. 2010; Gilroy et al. 2014). Sir Michael Berridge played an important role in identifying the role of inositol 1,4,5-trisphosphate (IP_3) in Ca^{2+} release insect salivary gland and established Ca^{2+} signaling as a paradigm (Bootman et al. 2020). Although Ca^{2+} fulfills the role of a macronutrient as well as a signaling molecule, it can also be cytotoxic in the cell in higher concentration (Hepler and Wayne 1985). It precipitates phosphate and as mentioned in the previous paragraph can compete with Mg^{2+} for enzyme binding site thus creating problems with biochemical reactions. Therefore, Ca^{2+} concentration in the cytosol needs to be kept at a very low level (Case et al. 2007). It is thought that this led to the genesis of the Ca^{2+} signaling (Sanders et al. 1999; Case et al. 2007). The cell achieves this low cytosolic Ca^{2+} levels through homeostatic machinery involving several Ca^{2+} transport elements (discussed in Chap. 2). There are certain properties of Ca^{2+} that makes it an ideal second messenger. Carafoli and Krebs have cited five properties in favor of Ca^{2+}- (1) the valency (coordination number is 6–8 but higher is possible *i.e.*, flexible coordination); (2) the ionic radius (99 pm compared to 65 pm for Mg^{2+}); (3) the polarizability, (it usually forms a pentagonal bipyramidal geometry in nature with binding affinity for carboxylate oxygen commonly found in amino acids); (4) the hydration energy (*i.e.*, the ease with which the water molecules are stripped off the metal –it is tougher to strip off water molecules from Mg^{2+} compared to Ca^{2+}); and (5) the radius of the hydrated metal ion, which determines the charge density (Hepler and Wayne 1985; Case et al. 2007; Carafoli and Krebs 2016). These properties allowed Ca^{2+} to rapidly bind and unbind from cellular sensor proteins (called Ca^{2+} binding proteins (CaBP), discussed in detail in Chap. 4). The interplay of the Ca^{2+} homeostasis and the basic biochemistry gives rise to another concept – the 'Ca^{2+} signature'. This hypothesis was forwarded by Sir Michael Berridge and was subsequently adapted for plants by Hetherington and colleagues (Berridge 1997; McAinsh et al. 1997). Ca^{2+} signature hypothesis led to our current understanding of Ca^{2+} signaling and is discussed in Chap. 3. This concept integrates the Ca^{2+} transport elements and CaBPs to stimulus-response-coupling.

Difference Between Animal and Plant Ca²⁺ Signaling- from the Perspective of Evolution

The first living organism (a cell-bound with membranes) originated in the ocean (Muller 1996; Verkhratsky and Parpura 2014). It had to maintain a proper cellular balance of ions, especially for Ca^{2+} (for the reasons already mentioned) and developed a Ca^{2+} homeostasis system (Case et al. 2007; Plattner and Verkhratsky 2015a). As the cell evolved, it needed some sort of a signaling mechanism that it could use to communicate and modulate its cellular responses- in short, it needed a communication system (Carafoli and Krebs 2016). Ca^{2+} that was thrown out of the cell for its cytotoxicity now became the molecule of choice because of its advantageous biochemical properties (discussed in the previous paragraph). Bacterial system (prokaryotes) possess efficient Ca^{2+} homeostasis system (the transport elements for pushing Ca^{2+} in and out of the cell) and CaBPs indicating the existence of Ca^{2+} signaling machinery (Dominguez et al. 2015; Plattner and Verkhratsky 2015b). Ca^{2+} signaling machinery (especially the homeostasis elements) is also observed in ancestral protists, which are probably the ancestors of modern eukaryotes (animals, fungi and plants) (Cai et al. 2015).

Since divergence occurred from this point, it is obvious that from this point some differences would be observed in the animal and plant Ca^{2+} signaling components. The basic principles of homeostasis and signaling remained the same, and the difference majorly occurred in the players involved (Edel et al. 2017). The evolution of plant Ca^{2+} homeostasis system was probably modulated to achieve terrestrial (land) colonization. There are certain properties in plants, which highly influenced this event. First, the land plants do not have four domain voltage-dependent Ca^{2+} channel (VDCCs), typically present in animals. Second, they lack (or fail to make use of) cyclic nucleotide-based second messenger system that is a potent signaling system in animals and crosstalk efficiently with Ca^{2+} signaling (in animals). The cyclic nucleotide signaling systems presence in plants is still controversial as the typical phosphodiesterase required are still unidentified (discussed more in Chap. 7).

Thirdly, in animal Na^+ ions is used for energizing the transport whereas plants use H^+ (Edel et al. 2017). Probably these facts (taken together with other yet unknown reasons) shaped the Ca^{2+} signaling mechanism of plants.

Based on the current genomic studies, plants have lost the Ca^{2+} influx elements such as animal homologues of VDCCs, inositol-1,4,5-triphosphate (IP$_3$) receptors, transient receptor potential (TRPs), purigenic receptors (P2XRs), and Cysteine loop channels (Cys-loops) during evolution (though some of them are present in chlorophytes and charophytes) (Verret et al. 2010; Edel and Kudla 2015). These are replaced by different influx transport elements discussed in Chap. 2. Some of them are exclusive to plants (like mechanosensitive channels) or are highly enriched in plant genome (like 20 cyclic nucleotide-gated channels in Arabidopsis versus 6 in *Homo sapiens*) (Verret et al. 2010). Plants have only one part (at a time) of the stromal interaction molecules (STIMs)-Orai that help in the formation of channels that release Ca^{2+} from the endoplasmic reticulum (ER) to the cytosol (Edel et al. 2017).

The CaBPs in plants are also very different and have less variation from their animal counterpart (Edel et al. 2017). The Calcineurin B-like (CBL) and CBL-interacting protein kinase (CIPK) represent a paradigm shift in signal transduction architecture of plants (Pandey 2008; Luan 2009; Pandey et al. 2014). This homologue of animal calcineurin (a phosphatase) is a kinase-based module indicating plants have probably adopted phosphorylation as a means for transducing Ca^{2+} signal. Similarly, Ca^{2+} dependent protein kinase (CDPKs) and CaM-like (CML) are absent in animals (at least in higher animals) (Sanyal et al. 2019). Also, plants have different Ca^{2+} extrusion transport elements. The cation/H^+ exchanger (CAX are present in plants but absent in *Homo sapiens* (present in fungi) (Verret et al. 2010). Similarly higher plants lack the K^+-independent Na^+/Ca^{2+} exchangers (NCXs) and K^+-dependent Na^+/Ca^{2+} exchanger (NCKX) (Cai et al. 2015).

Even though there are significant differences in the Ca^{2+} signaling toolkit (the homeostasis elements and CaBPs) between the animals and plants but plants do not appear to be at a disadvantaged position. Plants can still display many Ca^{2+} signaling attributes that are animal-specific. Despite the absence of animal IP_3 receptor homologue (also animal cADPR receptor homologue and NAADP receptor homologue), plants can still exhibit Ca^{2+} induced Ca^{2+} release (CICR) (discussed in Chap. 2) (Edel et al. 2017). Similarly, the action potential (AP) in plants may be generated in a different manner using Cl^- ions since the animal-like VDCC channels are absent in plants (Edel et al. 2017). Also, probably the absence of an animal like second messenger system is compensated by other hormone-regulated pathways with which Ca^{2+} signaling pathway crosstalk extensively. And finally, the CaBPs of plants have increased their number (each member is part of a large gene family that comprises of many proteins) so that they can efficiently cater to different stimulus. This versatility of the Ca^{2+} signaling toolkit in plants made plant biologist hypothesize that plant Ca^{2+} signaling toolkit is a robust system that serves the plant efficiently (Edel et al. 2017). Figure 1.2 provides a brief picture of components involved in plant and animal Ca^{2+} signaling.

Conclusion

Ca^{2+} performs a dual function in the cell- as a nutrient and a signaling molecule. Its role of a signaling molecule was shaped through evolution. "Homeostasis" and "Ca^{2+} signature" are two very important terms that shape the Ca^{2+} signaling events. The plant Ca^{2+} transport elements play a significant role in shaping these two phenomena. In the subsequent chapters, we look into these transport elements with greater detail.

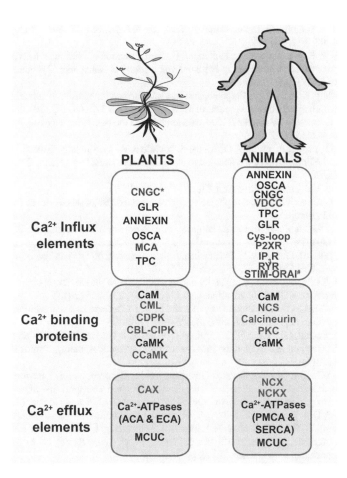

Fig. 1.2 The difference between the players involved in Ca²⁺ signal transduction in plants and animals. The comparison is majorly based on *Arabidopsis thaliana* (a model plant) and *Homo sapiens*. The unique players (on either side) are marked in red. Structure of plant CNGC differ from animal CNGC and hence is marked with an asterisk. Plants do not possess the animal STIM-ORAI together. Animal Ca²⁺-binding proteins are more in number, and only few are shown in the figure. In the subsequent chapters, we discuss more on the plant influx transport elements cyclic-nucleotide gated channels (CNGC), glutamate receptors (GLR), Annexin, Mechanosensitive channel (MCA), Two-pore channels and plant efflux transport elements Ca²⁺/H⁺ exchangers (CAX) and Ca²⁺ ATPases. The mitochondrial uniporter MCUC are discussed in Chap. 3. The plant Ca²⁺-binding proteins are discussed in Chap. 4. CCaMK and CaMK are not found in *Arabidopsis thaliana* but are included to show the CaBPs present in plants

References

M.J. Berridge, The AM and FM of calcium signalling. Nature **386**, 759–760 (1997)

M.J. Berridge, P. Lipp, M.D. Bootman, The versatility and universality of calcium signalling. Nat. Rev. Mol. Cell Biol. **1**, 11–21 (2000)

M. Bootman, A. Galione, C. Taylor, Professor Sir Michael Berridge FRS (1938–2020). Curr. Biol. **30**, R374–R376 (2020)

M. Broadley, P. Brown, I. Cakmak, Z. Rengel, F. Zhao, Function of Nutrients: Micronutrients, in *Marschner's Mineral Nutrition of Higher Plants*, ed. by P. Marschner, 3rd edn., (Academic, New York, 2012), pp. 191–248

X. Cai, X. Wang, S. Patel, D.E. Clapham, Insights into the early evolution of animal calcium signaling machinery: A unicellular point of view. Cell Calcium **57**, 166–173 (2015)

E. Carafoli, J. Krebs, Why calcium? How calcium became the best communicator. J. Biol. Chem. **291**, 20849–20857 (2016)

R.M. Case, D. Eisner, A. Gurney, O. Jones, S. Muallem, A. Verkhratsky, Evolution of calcium homeostasis: From birth of the first cell to an omnipresent signalling system. Cell Calcium **42**, 345–350 (2007)

D.E. Clapham, Calcium signaling. Cell **131**, 1047–1058 (2007)

S. Conn, M. Gilliham, Comparative physiology of elemental distributions in plants. Ann. Bot. **105**, 1081–1102 (2010)

A.N. Dodd, J. Kudla, D. Sanders, The language of calcium signaling. Annu. Rev. Plant Biol. **61**, 593–620 (2010)

D.C. Dominguez, M. Guragain, M. Patrauchan, Calcium binding proteins and calcium signaling in prokaryotes. Cell Calcium **57**, 151–165 (2015)

K.H. Edel, J. Kudla, Increasing complexity and versatility: How the calcium signaling toolkit was shaped during plant land colonization. Cell Calcium **57**, 231–246 (2015)

K.H. Edel, E. Marchadier, C. Brownlee, J. Kudla, A.M. Hetherington, The evolution of calcium-based signalling in plants. Curr. Biol. **27**, R667–r679 (2017)

S. Gilroy, N. Suzuki, G. Miller, W.G. Choi, M. Toyota, A.R. Devireddy, R. Mittler, A tidal wave of signals: Calcium and ROS at the forefront of rapid systemic signaling. Trends Plant Sci. **19**, 623–630 (2014)

A. González-Fontes, M.T. Navarro-Gochicoa, C.J. Ceacero, M.B. Herrera-Rodríguez, J.J. Camacho-Cristóbal, J. Rexach, Understanding Calcium Transport and Signaling, and Its Use Efficiency in Vascular Plants, in *Plant Macronutrient Use Efficiency*, ed. by M. A. Hossain, T. Kamiya, D. J. Burritt, L.-S. P. Tran, T. Fujiwara, (Academic, Oxford, 2017), pp. 165–180

M. Hawkesford, W. Horst, T. Kichey, H. Lambers, J. Schjoerring, I.S. Møller, P. White, Functions of Macronutrients, in *Marschner's Mineral Nutrition of Higher Plants*, ed. by P. Marschner, 3rd edn., (Academic, New York, 2012), pp. 135–189

P.K. Hepler, R.O. Wayne, Calcium and plant development. Ann. Rev. Plant Physiol. **36**, 397–439 (1985)

E. Kirkby, Introduction, definition and classification of nutrients, in *Marschner's Mineral Nutrition of Higher Plants*, ed. by P. Marschner, 3rd edn., (New York, Academic, 2012), pp. 3–5

J. Kudla, O. Batistic, K. Hashimoto, Calcium signals: The lead currency of plant information processing. Plant Cell **22**, 541–563 (2010)

S. Luan, The CBL-CIPK network in plant calcium signaling. Trends Plant Sci. **14**, 37–42 (2009)

F.J. Maathuis, Physiological functions of mineral macronutrients. Curr. Opin. Plant Biol. **12**, 250–258 (2009)

P. Marschner, Z. Rengel, Nutrient Availability in Soils, in *Marschner's Mineral Nutrition of Higher Plants*, ed. by P. Marschner, 3rd edn., (Academic, New York, 2012), pp. 315–330

M.R. McAinsh, C. Brownlee, A.M. Hetherington, Calcium ions as second messengers in guard cell signal transduction. Physiol. Plant. **100**, 16–29 (1997)

A.W. Muller, Hypothesis: The thermosynthesis model for the origin of life and the emergence of regulation by Ca^{2+}. Essays Biochem. **31**, 103–119 (1996)

G.K. Pandey, Emergence of a novel calcium signaling pathway in plants: CBL-CIPK signaling network. Physiol. Mol. Biol. Plants **14**, 51–68 (2008)

G.K. Pandey, P. Kanwar, A. Pandey, *Global Comparative Analysis of CBL-CIPK Gene Families in Plants* (Springer, New York, 2014)

H. Plattner, A. Verkhratsky, Evolution of calcium signalling. Cell Calcium **57**, 121–122 (2015a)

H. Plattner, A. Verkhratsky, The ancient roots of calcium signalling evolutionary tree. Cell Calcium **57**, 123–132 (2015b)

D. Sanders, C. Brownlee, J.F. Harper, Communicating with calcium. Plant Cell **11**, 691–706 (1999)

S.K. Sanyal, S. Mahiwal, G.K. Pandey, Calcium Signaling: A Communication Network that Regulates Cellular Processes, in *Sensory Biology of Plants*, ed. by S. Sopory, (Springer, Singapore, 2019), pp. 279–309

S. Stael, B. Wurzinger, A. Mair, N. Mehlmer, U.C. Vothknecht, M. Teige, Plant organellar calcium signalling: An emerging field. J. Exp. Bot. **63**, 1525–1542 (2012)

K. Thor, Calcium-nutrient and messenger. Front. Plant Sci. **10**, 440 (2019)

A. Trewavas, Le calcium, C'est la vie: Calcium makes waves. Plant Physiol. **120**, 1–6 (1999)

A.J. Trewavas, R. Malho, Ca^{2+} signalling in plant cells: The big network! Curr. Opin. Plant Biol. **1**, 428–433 (1998)

A. Verkhratsky, V. Parpura, Calcium signalling and calcium channels: Evolution and general principles. Eur. J. Pharmacol. **0**, 1–3 (2014)

F. Verret, G. Wheeler, A.R. Taylor, G. Farnham, C. Brownlee, Calcium channels in photosynthetic eukaryotes: Implications for evolution of calcium-based signalling. New Phytol. **187**, 23–43 (2010)

Chapter 2
Calcium Homeostasis, Reserves and Transport Elements in the Cell

Contents

Introduction to the Concept of Ca²⁺ Homeostasis

In the previous chapter, we briefly used the term Ca²⁺ homeostasis to explain how the cell manage to maintain low levels of Ca²⁺ to escape its negative impact. The cell must maintain a perfect balance for all cellular metabolites for its proper functioning. This is also true for Ca²⁺, which in excess can precipitate phosphates (Dodd et al. 2010). But as the cell employed Ca²⁺ as a second messenger, it would have been a waste of valuable resource if the cell would simply have dumped the excess Ca²⁺ away. So, it came up with a balanced strategy- it removed the excess cytosolic Ca²⁺ but stored it in nearby compartments where the Ca²⁺ could be brought back into the cytoplasm at a short notice. Any relevant stimuli would result in the release of Ca²⁺ from these stores, resulting in the transient rise of Ca²⁺ in the cytosol (Sanders et al. 2002; Hetherington and Brownlee 2004; McAinsh and Pittman 2009). The end of the stimuli would see the excess Ca²⁺ returning back to the stores. In between these two, even a separate phenomenon is known as "Ca²⁺ signature" occurs, that carries the message coded in a cytosolic Ca²⁺ transient (McAinsh and Pittman 2009; Sanyal et al. 2019). We will talk about this Ca²⁺ signature event more in Chap. 3.

Coming back to the sudden Ca²⁺ rise and resultant extrusion, the cell needed mechanisms that were fast in their job and placed in strategic locations. The cell have chosen the apoplast, endoplasmic reticulum (ER), vacuole (aka tonoplast) and Golgi as major Ca²⁺ stores (Stael et al. 2012; Costa et al. 2018). The cell also uses mitochondria and chloroplast, but we will see later in Chap. 3, that these have their separate Ca²⁺ signature events. This is also true for the nucleus and the peroxi-

© The Editor(s) (if applicable) and The Author(s), under exclusive license to
Springer Nature Switzerland AG 2021
G. K. Pandey, S. K. Sanyal, *Functional Dissection of Calcium Homeostasis and Transport Machinery in Plants*, SpringerBriefs in Plant Science,
https://doi.org/10.1007/978-3-030-58502-0_2

somes. For simplifying the concept- we consider the apoplast, ER, vacuole and Golgi as stores and the -rest (mitochondria, chloroplast, nucleus and peroxisomes) as stores cum playfields of Ca^{2+} signaling. The cell has equipped both with certain mechanisms- the Ca^{2+} influx mechanism (to pump in Ca^{2+} at the beginning of stimuli) and -Ca^{2+} efflux mechanism (to pump out Ca^{2+} at the end of stimuli). These entire set- the stores and the Ca^{2+} influx/efflux mechanism work in synergy to maintain a nontoxic cytosolic resting Ca^{2+} level (about 100 nM) and thus constitute the Ca^{2+} homeostasis (Costa et al. 2018). We will discuss more about the stores and stores cum playfields (only their storage capacity) in the chapter and very briefly mention the influx and efflux mechanisms, which we are discussing in more detail in the subsequent chapters.

The Ca^{2+} Stores- Apoplast, Vacuole, ER, and Golgi

The apoplast is the first Ca^{2+} store of a cell as well as the first compartment that perceives a stimulus (Gao et al. 2004). The total Ca^{2+} present in an organelle (or store) is mostly kept in bound form and only some amount is free for signaling (Stael et al. 2012). The term total and free Ca^{2+} used in this chapter, is the demarcation between these two forms. Apoplast receives Ca^{2+} through water transpiration and it is estimated that it can store Ca^{2+} in the range of 1 mM (total) of which about 0.33 mM is free resting Ca^{2+} (Stael et al. 2012). The Ca^{2+} in the apoplast is majorly stored in bound form with pectin and oxalates and are important in maintenance of plant cell wall rigidity (by binding to pectate) and also involved in proper stomatal movement (Hepler 2005). Several research groups have shown that when they block Ca^{2+} (or other ion extrusion systems) from the plasma membrane (PM), the Ca^{2+} transient of the cytoplasm is severely perturbed (Knight et al. 1996; Lamotte et al. 2004; Ali et al. 2007; Navazio et al. 2007). So this proves that the apoplast also acts as one of the primary sources for generating cytosolic Ca^{2+} rise in response to any stimulus. The other cellular stores also influence the apoplastic Ca^{2+} store, as it has been proved that knockout of important Ca^{2+} transport elements (cyclic nucleotide-gated channels (CNGC) and cation/proton exchangers (CAX) lead to the over-accumulation of apoplastic Ca^{2+} (Conn et al. 2011; Wang et al. 2017). It indicates that there is a clear transport pathway defined in plants, and disrupting this pathway leads to unwanted accumulation of Ca^{2+} in cytoplasm.

The vacuole is the largest storehouse of Ca^{2+} in the plant cell (Stael et al. 2012; Costa et al. 2018). Along with Ca^{2+}, it stores several other primary and secondary metabolites (Kruger and Schumacher 2018; Shimada et al. 2018). Like in the case of apoplast, majority of the Ca^{2+} in the vacuole is kept bound with malate, citrate and isocitrate (Stael et al. 2012). The vacuolar Ca^{2+} may also be kept bound with some proteins like radish vacuolar Ca^{2+} binding protein (RVCaB) (Yuasa and Maeshima 2001). As such out of a probable (50–80 mM) of Ca^{2+} in the vacuole only about 0.2–5 mM is available in free form (Stael et al. 2012). The vacuoles thus contribute in generating cytosolic Ca^{2+} transients due to its high free Ca^{2+} availability (highest among the stores) (Stael et al. 2012). However, Ca^{2+} induced Ca^{2+} release (CICR) is believed to be fueled by the vacuole. In the animal cell, it is an established concept that stimuli can activate the Ryanodine receptors (majorly located in the cell organelles) to mediate Ca^{2+} release in the cytosol (Dodd et al. 2010). The receptors are activated by Ca^{2+} transients in the cytosol and thus activate the CICR. The CICR can amplify a signal by enhancing the amount of Ca^{2+} released in the cytosol. The plant vacuole and ER are hypothesized to take part in the CICR phenomenon (White and Broadley 2003; Dodd et al. 2010). However, the mode of generating the CICR may be different as most of the typical animal receptors involved in CICR are absent in plants (Dodd et al. 2010). The slow activating vacuolar channel (SV channels) is now considered to be Two-pore channel 1 (TPC1) that releases Ca^{2+} from the vacuole to elicit CICR in plants (Dodd et al. 2010). There can be other pathways that crosstalk with the Ca^{2+} signaling event and modulate CICR in plants. We talk more about CICR at the end of this chapter.

The next storehouse in the plant cell is the ER. The actual concentration of Ca^{2+} stored in plant ER is unknown, and so the animal value (2 mM total and about 0.5 mM free) is taken as the current working point (Stael et al. 2012). Like the other organelles, Ca^{2+} in ER is also kept in bound form by calreticulin (Christensen et al. 2010). The specialized structure and arrangement of the ER also allow it to perform inter-organellar exchanges (Costa et al. 2018). The best example for this is perhaps the direct exchange of Ca^{2+} between the ER and the nucleus to generate a nucleus specific Ca^{2+} transient (Charpentier 2018). There are increasing pieces of evidence that suggest that the ER and plastids also engage in metabolite exchange (and probably even Ca^{2+}) through specialized stroma filled tubules known as stromules (Schattat et al. 2011). The ER and PM have similar contact sites that can be a hub to form a Ca^{2+} transient (Bayer et al. 2017). The ER and mitochondria contact can shape the cytosolic Ca^{2+} transient in animals, but concrete proof for similar association in plants is lacking (Costa et al. 2018). Figure 2.1 depicts a hypothetical model of plant vacuole and ER shaping the CICR event in plants.

Compared to the other three stores (apoplast, vacuole and ER) information on Golgi as a Ca^{2+} store has not been reported very extensively. The plant Golgi has a free Ca^{2+} concentration of 0.7 mM and calreticulin keeps the Ca^{2+} in the bound form like in the ER (Costa et al. 2018).

The Stores Cum Playfields-Mitochondria, Chloroplast, Nucleus and Peroxisomes

The chloroplast, mitochondria, nucleus and peroxisomes are now considered to have separate Ca^{2+} signaling events (than the cytoplasm). However, these organelles also store Ca^{2+} (Stael et al. 2012; Costa et al. 2018). Among them, chloroplast is the largest store with about 15 mM of total Ca^{2+} which remains mostly bound to the thylakoid lumen and the stroma has about 150 nM free Ca^{2+} (Stael et al. 2012;

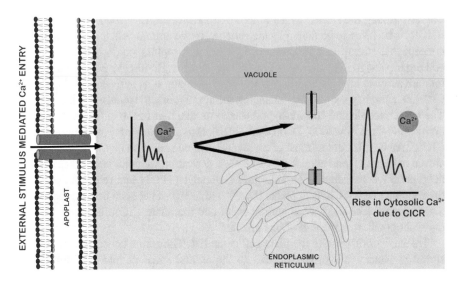

Fig. 2.1 The Ca^{2+}-induced Ca^{2+} release (CICR) in a plant cell. A hypothetical model describing the probable cytosolic Ca^{2+} transient increase in the plant cell through CICR. Any external stimulus would enhance the cytosolic Ca^{2+}, followed by this Ca^{2+} (and other second messengers) regulating certain transport elements of ER and vacuole to release more Ca^{2+} in the cytoplasm. This entire mechanism would increase the cytosolic Ca^{2+} thus amplifying the Ca^{2+} transient. Description of the event and the involved players are mentioned in the text

Nomura and Shiina 2014). The mitochondrial matrix has about 600 nM of free Ca^{2+} (Costa et al. 2018). The amyloplast (plastid stroma) has about 100 nM, nucleus (about 100 nM) and peroxisomes (about 2 µM) (Stael et al. 2012; Costa et al. 2018). We discuss in detail about them in Chap. 3.

The Transport Element- "the Actors" Involved in Ca^{2+} Homeostasis

The research on ion channels had started when researchers were focusing to study electrical impulses in animals and plants (Hedrich 2012). The breakthrough in this study was achieved with the invention of the patch-clamp technique (Hedrich 2012). This technique (aka electrophysiology) allowed the measurement of single-channel current analyses and characterization of ion channels in higher plants (Hedrich 2012). However, one major problem, which still exists to date, is the absence of animal VDCCs homologues in the plant genome (Dodd et al. 2010). Nevertheless, the PM localized depolarization activated Ca^{2+} channel (DACC), hyperpolarization-activated Ca^{2+} channel (HACC) and voltage insensitive Ca^{2+} channel (VICC) were identified depending on their electrophysiological properties (Dodd et al. 2010). They were also classified as non-selective cation channel (NSCC) since they are permeable to a wide range of cations (the classification stays the same-DA-NSCC, HA-NSCC and VI-NSCC) (Demidchik and Maathuis 2007). These channels are not strictly selective to Ca^{2+} as the name suggests. Their physiological properties are also different- DACC allows transient Ca^{2+} influx, HACC allowed large Ca^{2+} influx and VICC allowed Ca^{2+} influx under weak (or physiological) voltage to maintain the resting Ca^{2+} concentration in the cytosol (Sanyal et al. 2019). The DACC and VICC were observed in roots, xylem, guard cells among others. The HACC is observed more in the root hairs and cells in the elongation zone (Demidchik and Maathuis 2007). The similar voltage-dependent activity was identified at vacuole and classified as SV and fast vacuolar (FV) channel (Demidchik and Maathuis 2007). The FV channel is permeable to potassium (K). The SV channel, as already mentioned, was later identified as TPC1 (and is the only channel with confirmed molecular identity). Similarly in the chloroplast another channel identified through electrophysiology, the voltage-dependent fast-activating cation channel (FACC) (Pottosin et al. 2005). The authors have hypothesized that the FACC may be GLR3.4 localized to the chloroplast (Pottosin and Shabala 2016).

Besides these channels identified by electrophysiology, there is another group of channels that await molecular identification. It is known that intracellular second messengers like IP$_3$ (inositol 1,4,5-triphosphate) and cADPR (cADP ribose) activate distinct Ca^{2+} release channels in animals. These ligands also release Ca^{2+} from VM and ER thus modulating the CICR in plants (White 2000; White and Broadley 2003). But again the mammalian homologues of IP$_3$ and cADPR receptors are not identified yet in the plant genome (Dodd et al. 2010).

However, several other candidates have been now identified that have a presence in the plant genome. The candidates that are involved in Ca^{2+} influx are indicated first. The ligand-gated channels are important candidates involved in Ca^{2+} entry into the cytosol. The cyclic nucleotide-gated channels (CNGC) and the glutamate receptors channels (GLRs) are members of this class (Dodd et al. 2010; DeFalco et al. 2016; Jha et al. 2016). Then there are annexins, which are membrane trafficking proteins but can form voltage-gated cation channel (Kudla et al. 2010). There is also a group of mechanosensitive channels classified as- mechanosensitive-like channel (MSL), Mid1-complementing channels (MCA), hyperosmolarity-induced [Ca^{2+}]$_{cyt}$ increase' channels (OSCA), and Piezo channels (Hedrich 2012; Yuan et al. 2014; Hamilton et al. 2015).

The Ca^{2+} efflux machinery works to maintain basal level of Ca^{2+} in the cytoplasm by extruding it to the Ca^{2+} stores. The plant cell has two broad groups working to efflux Ca^{2+}. The first group is the Ca^{2+}/cation antiporter (CaCA) superfamily and second is the P-type Ca^{2+} ATPases (Yadav et al. 2012; Singh et al. 2014; Sanyal et al. 2019). The former immediately lower Ca^{2+} level in the cell and the later maintains resting Ca^{2+} levels. The H$^+$/Cation exchanger (CAX) (a member of CaCA superfamily) use electrochemical gradient for Ca^{2+} transport and are low affinity high capacity Ca^{2+} transporters (Demidchik et al. 2018). The Ca^{2+} ATPases (ER-Type Ca^{2+}-ATPases (ECA) and autoinhibited Ca^{2+}-ATPases (ACA)) use ATP hydrolysis to power Ca^{2+} transport and have low capacity but high affinity Ca^{2+} (Demidchik et al. 2018). We will discuss the main influx and efflux channels in detail through

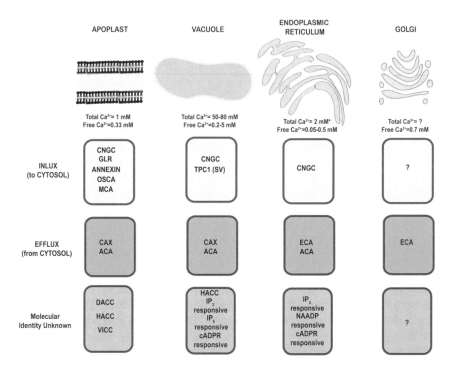

Fig. 2.2 Transport elements present in the Ca²⁺ "stores" of a plant cell. The Ca²⁺ transport elements (influx and efflux) of the different Ca²⁺ stores (apoplast, ER, vacuole, Golgi) are mentioned along with the probable Ca²⁺ concentration in each compartment (values according to (Stael et al. 2012)). The ER values are from an animal cell. Some of the players, whose molecular identity is unknown (discussed in the text) are mentioned at their putative location according to (White and Broadley 2003; Dodd et al. 2010). The elements that fall under the category of "player cum stores" are mentioned in a separate figure in Chap. 3

Chaps. 5, 6, 7, 8, and 9. Figure 2.2 denotes an overview of the organelles/subcellular locations where these elements (putatively) operate.

Conclusion

Ca²⁺ homeostasis is an important event in the plant cell. This controls the generation of a more complex physiological phenomenon "Ca²⁺ signature". Although the electrophysiology and ligand-based response approach have shown that there are different classes of Ca²⁺ transporter elements in the cell, their molecular identification is still pending. Further research should identify these unknown candidates.

References

R. Ali, W. Ma, F. Lemtiri-Chlieh, D. Tsaltas, Q. Leng, S. von Bodman, G.A. Berkowitz, Death don't have no mercy and neither does calcium: Arabidopsis CYCLIC NUCLEOTIDE GATED CHANNEL2 and innate immunity. Plant Cell **19**, 1081–1095 (2007)

E.M. Bayer, I. Sparkes, S. Vanneste, A. Rosado, From shaping organelles to signalling platforms: The emerging functions of plant ER-PM contact sites. Curr. Opin. Plant Biol. **40**, 89–96 (2017)

M. Charpentier, Calcium signals in the plant nucleus: Origin and function. J. Exp. Bot **69**, 4165–4173 (2018)

A. Christensen, K. Svensson, L. Thelin, W. Zhang, N. Tintor, D. Prins, N. Funke, M. Michalak, P. Schulze-Lefert, Y. Saijo, M. Sommarin, S. Widell, S. Persson, Higher plant calreticulins have acquired specialized functions in Arabidopsis. PLoS One **5**, e11342 (2010)

S.J. Conn, M. Gilliham, A. Athman, A.W. Schreiber, U. Baumann, I. Moller, N.H. Cheng, M.A. Stancombe, K.D. Hirschi, A.A. Webb, R. Burton, B.N. Kaiser, S.D. Tyerman, R.A. Leigh, Cell-specific vacuolar calcium storage mediated by CAX1 regulates apoplastic calcium concentration, gas exchange, and plant productivity in Arabidopsis. Plant Cell **23**, 240–257 (2011)

A. Costa, L. Navazio, I. Szabo, The contribution of organelles to plant intracellular calcium signaling. J. Exp. Bot. **69**, 4175–4193 (2018)

T.A. DeFalco, W. Moeder, K. Yoshioka, Opening the gates: Insights into cyclic nucleotide-Gated Channel-mediated signaling. Trends Plant Sci. **21**, 903–906 (2016)

V. Demidchik, F.J. Maathuis, Physiological roles of nonselective cation channels in plants: From salt stress to signalling and development. New Phytol. **175**, 387–404 (2007)

V. Demidchik, S. Shabala, S. Isayenkov, T.A. Cuin, I. Pottosin, Calcium transport across plant membranes: Mechanisms and functions. New Phytol. **220**, 49–69 (2018)

A.N. Dodd, J. Kudla, D. Sanders, The language of calcium signaling. Annu. Rev. Plant Biol. **61**, 593–620 (2010)

D. Gao, M.R. Knight, A.J. Trewavas, B. Sattelmacher, C. Plieth, Self-reporting Arabidopsis expressing pH and [Ca2+] indicators unveil ion dynamics in the cytoplasm and in the apoplast under abiotic stress. Plant Physiol. **134**, 898–908 (2004)

E.S. Hamilton, A.M. Schlegel, E.S. Haswell, United in diversity: Mechanosensitive ion channels in plants. Annu. Rev. Plant Biol. **66**, 113–137 (2015)

R. Hedrich, Ion channels in plants. Physiol. Rev. **92**, 1777–1811 (2012)

P.K. Hepler, Calcium: A central regulator of plant growth and development. Plant Cell **17**, 2142–2155 (2005)

A.M. Hetherington, C. Brownlee, The generation of Ca^{2+} signals in plants. Annu. Rev. Plant Biol. **55**, 401–427 (2004)

S.K. Jha, M. Sharma, G.K. Pandey, Role of cyclic nucleotide gated channels in stress management in Plants. Curr. Genomics **17**, 315–329 (2016)

H. Knight, A.J. Trewavas, M.R. Knight, Cold calcium signaling in Arabidopsis involves two cellular pools and a change in calcium signature after acclimation. Plant Cell **8**, 489–503 (1996)

F. Kruger, K. Schumacher, Pumping up the volume – Vacuole biogenesis in Arabidopsis thaliana. Semin. Cell Dev. Biol. **80**, 106–112 (2018)

J. Kudla, O. Batistic, K. Hashimoto, Calcium signals: The lead currency of plant information processing. Plant Cell **22**, 541–563 (2010)

O. Lamotte, K. Gould, D. Lecourieux, A. Sequeira-Legrand, A. Lebrun-Garcia, J. Durner, A. Pugin, D. Wendehenne, Analysis of nitric oxide signaling functions in tobacco cells challenged by the elicitor cryptogein. Plant Physiol. **135**, 516–529 (2004)

M.R. McAinsh, J.K. Pittman, Shaping the calcium signature. New Phytol. **181**, 275–294 (2009)

L. Navazio, R. Moscatiello, A. Genre, M. Novero, B. Baldan, P. Bonfante, P. Mariani, A diffusible signal from arbuscular mycorrhizal fungi elicits a transient cytosolic calcium elevation in host plant cells. Plant Physiol. **144**, 673–681 (2007)

H. Nomura, T. Shiina, Calcium signaling in plant endosymbiotic organelles: Mechanism and role in physiology. Mol. Plant **7**, 1094–1104 (2014)

I. Pottosin, S. Shabala, Transport across chloroplast membranes: Optimizing photosynthesis for adverse environmental conditions. Mol. Plant **9**, 356–370 (2016)

I.I. Pottosin, M. Martinez-Estevez, O.R. Dobrovinskaya, J. Muniz, Regulation of the slow vacuolar channel by luminal potassium: Role of surface charge. J. Membr. Biol. **205**, 103–111 (2005)

D. Sanders, J. Pelloux, C. Brownlee, J.F. Harper, Calcium at the crossroads of signaling. Plant Cell **14**(Suppl), S401–S417 (2002)

S.K. Sanyal, S. Mahiwal, G.K. Pandey, Calcium Signaling: A Communication Network that Regulates Cellular Processes, in *Sensory Biology of Plants*, ed. by S. Sopory, (Springer, Singapore, 2019), pp. 279–309

M. Schattat, K. Barton, B. Baudisch, R.B. Klosgen, J. Mathur, Plastid stromule branching coincides with contiguous endoplasmic reticulum dynamics. Plant Physiol. **155**, 1667–1677 (2011)

T. Shimada, J. Takagi, T. Ichino, M. Shirakawa, I. Hara-Nishimura, Plant Vacuoles. Annu. Rev. Plant Biol. **69**, 123–145 (2018)

A. Singh, P. Kanwar, A.K. Yadav, M. Mishra, S.K. Jha, V. Baranwal, A. Pandey, S. Kapoor, A.K. Tyagi, G.K. Pandey, Genome-wide expressional and functional analysis of calcium transport elements during abiotic stress and development in rice. FEBS J. **281**, 894–915 (2014)

S. Stael, B. Wurzinger, A. Mair, N. Mehlmer, U.C. Vothknecht, M. Teige, Plant organellar calcium signalling: An emerging field. J. Exp. Bot. **63**, 1525–1542 (2012)

Y. Wang, Y. Kang, C. Ma, R. Miao, C. Wu, Y. Long, T. Ge, Z. Wu, X. Hou, J. Zhang, Z. Qi, CNGC2 is a Ca2+ influx channel that prevents accumulation of Apoplastic Ca2+ in the leaf. Plant Physiol. **173**, 1342–1354 (2017)

P.J. White, Calcium channels in higher plants. Biochim. Biophys. Acta **1465**, 171–189 (2000)

P.J. White, M.R. Broadley, Calcium in plants. Ann. Bot. **92**, 487–511 (2003)

A.K. Yadav, A. Pandey, G.K. Pandey, Calcium homeostasis: Role of CAXs transporters in plant signaling. Plant Stress **6**, 60–69 (2012)

F. Yuan, H. Yang, Y. Xue, D. Kong, R. Ye, C. Li, J. Zhang, L. Theprungsirikul, T. Shrift, B. Krichilsky, D.M. Johnson, G.B. Swift, Y. He, J.N. Siedow, Z.M. Pei, OSCA1 mediates osmotic-stress-evoked Ca2+ increases vital for osmosensing in Arabidopsis. Nature **514**, 367–371 (2014)

K. Yuasa, M. Maeshima, Organ specificity of a vacuolar Ca2+−binding protein RVCaB in radish and its expression under Ca2+−deficient conditions. Plant Mol. Biol. **47**, 633–640 (2001)

Chapter 3
The Concept of Calcium Signature and Its Involvement in Other Signaling Pathways

Contents

Calcium Signature –What and Why?

Among the different criteria that define a signaling molecule, one major requirement is that its concentration in the cell must be regulated, i.e., it should be able to jump from a very low concentration to a high concentration in the cell. As we have already seen in the previous chapter, the Ca^{2+} concentration in the cell can be regulated through homeostatic machinery. Hetherington and colleagues put forward the concept of "Ca^{2+} signature" to explain this phenomenon in plants (McAinsh et al. 1997; Kudla et al. 2010). They predicted that in response to different stimuli, the cellular Ca^{2+} concentration raises differing in their amplitude, time (duration), and space (McAinsh and Hetherington, 1998). According to this hypothesis, a stimulus would result in the upregulation of cytosolic Ca^{2+} (also mentioned in the previous chapter for understanding this concept). Cytosol is still the major area where the Ca^{2+} signature originates after (any) signal perturbs the cell (Dodd et al. 2010; Stael et al. 2012). This increase in the Ca^{2+} in the cytosol is rapid and generated in response to a variety of stimuli (Sanyal et al. 2019). The stimuli are then decoded by the large array of Ca^{2+} binding proteins (CaBP) available in the cell (discussed a little here in this chapter and in more detail in next chapter).

Current research has proved that cellular compartments like mitochondria, peroxisomes, chloroplast and nucleus could also generate their separate Ca^{2+} signatures, which determine the signal transduction events inside these organelles (Stael

© The Editor(s) (if applicable) and The Author(s), under exclusive license to
Springer Nature Switzerland AG 2021
G. K. Pandey, S. K. Sanyal, *Functional Dissection of Calcium Homeostasis and Transport Machinery in Plants*, SpringerBriefs in Plant Science,
https://doi.org/10.1007/978-3-030-58502-0_3

et al. 2012; Wagner et al. 2016; Charpentier 2018; Costa et al. 2018; Navazio et al. 2020). An important question here is- do the cytosolic Ca^{2+} signature crosstalk with organeller Ca^{2+} signatures and help in their generation? Or, the organeller Ca^{2+} signatures can bypass the cytosolic ones and can be generated on their own? There is a possibility that some cytosolic players can help in the generation of the organellar signals and there is an equal chance that they can have a separate pathway as well.

Ca^{2+} Signaling in Organelles

Chloroplast

Ca^{2+} plays an important role in modulating the important physiological reactions inside the chloroplast (Stael et al. 2012; Navazio et al. 2020). But the Ca^{2+} concentration needs tight regulation as it can hinder the normal functioning of the chloroplast (a similar situation of precipitating phosphates inside the chloroplast) (Stael et al. 2012). So there must be a mechanism that regulates chloroplastic Ca^{2+} levels. The major changes in the Ca^{2+} spike in the chloroplast is seen when the plants make a transition from dark to light (Kreimer et al. 1985; Roh et al. 1998). It is believed that the excess Ca^{2+} is kept in a chloroplastic store, which could be the thylakoid membrane and/or an unidentified store inside the chloroplast (Stael et al. 2012). There is also a possibility that phosphorylated protein residues of the thylakoid can act as buffers to bind excess Ca^{2+}. The important cog in the wheel is the elements that regulate the in and out passage of Ca^{2+} into the chloroplast. The membrane potential in the chloroplast envelope helps in Ca^{2+} uptake inside the chloroplast, but this transporter (probably a uniporter) is yet to be identified (Kreimer et al. 1985; Miller and Sanders 1987; Roh et al. 1998). Recent studies have identified chloroplast mitochondrial Ca^{2+} uniporter (AtcMCU), AtGLR3.4 and AtGLR3.5 in chloroplast Ca^{2+} transport that are also involved in abiotic stress tolerance (Teardo et al. 2010; Teardo et al. 2011; Teardo et al. 2019). Besides these, two ATPases, AtACA1 and AtHMA1 are believed to be involved in Ca^{2+} transport inside the chloroplast (Huang et al. 1993; Moreno et al. 2008). There are some doubts with AtACA1 regarding its actual localization and absence of Ca^{2+} ATPase activity in the envelope of the chloroplast (Malmstrom et al. 1997; Dunkley et al. 2006; Stael et al. 2012). AtHMA1 also comes with its own sets of contradictions. Although it can show high-affinity Ca^{2+} transport and is blocked by thapsigargin (a Ca^{2+}-ATPase inhibitor), it was later related to Cu^{2+} and Zn^{2+} transport (Seigneurin-Berny et al. 2006; Kim et al. 2009). AtACA9, AtBICAT1 (PAM71-type manganese/ Ca^{2+} cation transporter), AtBICAT2, AtOSCA2.5 may also be involved in chloroplast Ca^{2+} transport (Schiott et al. 2004; Lee et al. 2011; Frank et al. 2019). Other interesting candidates are the integral membrane protein ALBINO3 (ALB3) and pea chloroplast Pisum-post-floral-specific gene 1 (PPF1) (a gibberellin induced gene involved in flower development) (Sundberg et al. 1997; Wang et al. 2003). At this moment not much is known about the probable extrusion mechanism present in the chloroplast.

CaBP play an important role in chloroplast function. Calmodulin (CaMs), TOC (translocon at the outer envelope of chloroplasts) and TIC (translocon at the inner envelope of chloroplasts) complexes play an important role in the import of nuclear-encoded chloroplast proteins (Chigri et al. 2005; Chigri et al. 2006; Stael et al. 2012). CaM also activates the NAD kinase, which can then mediate the conversion of NAD to NADP (Jarrett et al. 1982; Takahashi et al. 2006). Ca^{2+} dependent protein kinases (CDPKs (a.k.a CPKs)) (CPK20 and CPK31) are also localized to the chloroplast (Dammann et al. 2003; Behrens et al. 2013; Helm et al. 2014; Ji et al. 2017). Besides, chloroplast has a few novel EF-hand containing proteins present. The first is Ca^{2+}-activated RelA/SpoT homologous protein (CRSH) and responsible for the production of guanosine 5′-diphosphate 3′-diphosphate (ppGpp) (Masuda et al. 2008). The levels of ppGpp change after plant encounter different stress, however, it is still not clear if there is clear crosstalk between this molecule and Ca^{2+} to propagate stress signals (Stael et al. 2012). The second EF-hand containing chloroplastic protein is SAMT-like (AtSAMT), which is proposed to import S-adenosylmethionine (SAM) into the chloroplast (Stael et al. 2011). The thylakoid membrane-localized Ca^{2+} sensing receptor (CAS) proteins have Ca^{2+}-binding domain and are the main target of Ca^{2+}-induced phosphorylation in chloroplasts (Cutolo et al. 2019). CAS is involved in the generation of Ca^{2+} induced Ca^{2+} transients (in the cytosol) and Ca^{2+} uptake (in guard cells) (Stael et al. 2012). CAS can regulate stomatal closure in response to Ca^{2+} (Weinl et al. 2008).

Mitochondria

Mitochondria are actively involved in regulating many physiological aspects in the cell –e.g., they help in ATP generation, ROS generation and are involved in Ca^{2+} signaling (Wagner et al. 2016; Kanwar et al. 2020). The recent years have seen a surge in information regarding mitochondrial Ca^{2+} transport. One important concept about mitochondria is that it can act as store house for Ca^{2+}, and along with endoplasmic reticulum can generate Ca^{2+} microdomains to modulate Ca^{2+} signature (Pedriali et al. 2017). For subsequent Ca^{2+} intake, the first barrier is the mitochondrial outer membrane (OMM), and voltage-dependent anion channels (VDACs) are thought to be the chief patrons assisting in the Ca^{2+} import (at least in animals) (Kanwar et al. 2020). The role of plant VDACs in importing Ca^{2+} is still debated as there is no solid proof to argue in its favor. However, some pieces of evidence like Arabidopsis VDACs interaction with plant Calcineurin-B like (CBL) proteins indicate that they may be involved in Ca^{2+} import like their animal counterparts (Li et al. 2013).

Once Ca^{2+} has crossed into the inner mitochondrial space (IMS), it has to get through the inner mitochondrial membrane (IMM) to reach the mitochondrial matrix. It is now accepted that the mitochondrial calcium uniporter complex

(MCUC) play a major role in this transport (Wagner et al. 2016). The current mammalian model suggests that the mammalian MCUC consist of the pore-forming protein mitochondrial calcium uniporter (MCU), an MCU paralogue (MCUb), the essential MCU regulator (EMRE), the regulatory MICU proteins, and possibly, the mitochondrial calcium uniport regulator 1 (MCUR1) (Wagner et al. 2016). The MCU protein tetramerizes to form a functional pore that allows Ca^{2+} entry into the matrix. MCUb can interact with MCU (heteromerization) to regulate MCU (dominant-negative regulation) (De Stefani et al. 2011). Plants have reported the existence of MCU homologues in the genome but experimental validation for their role in Ca^{2+} transport is still awaited. Another regulator of the MCU is the pseudo-EF-hand containing MICU proteins that are present in the IMS. The MICU can activate (or inhibit) the MCU. For this MICU employs its different isoforms that are found in the mammalian cells (MICU1, MICU2 and MICU3) (Plovanich et al. 2013; Waldeck-Weiermair et al. 2015). Again like MCU, MICU homologues are also found in plants but unlike the animals, the plant MICU have three EF-hands, which indicate that the protein can function differently in plant system (Wagner et al. 2015). The EMRE serve as a bridge between MCU and MICU and it is believed that EMRE senses the Ca^{2+} concentration in the mitochondrial matrix (Sancak et al. 2013). The plants probably do not possess MCUb and EMRE regulators.

Besides the core MCU, several other candidates can have a putative role in Ca^{2+} transport. The Ca^{2+} binding mitochondrial carriers (CamC) category of proteins can be divided into two classes: aspartate/glutamate carriers (AGCs) and ATP/Pi carriers (APCs/SCaMCs/SLCs). Arabidopsis encodes three mitochondrial small Ca^{2+} binding mitochondrial carriers (SCamC) that bind to Ca^{2+} (Stael et al. 2011). The glutamate receptor 3.5 (GLR3.5) is another promising candidate and is speculated that it may play a role in Ca^{2+} uptake in response to developmental stimuli (Teardo et al. 2015). Once the Ca^{2+} crosses the IMM, it enters the matrix where it exists primarily as insoluble Ca^{2+}-phosphate precipitate and bound to other proteins and inorganic acids (Wagner et al. 2016). This allows the mitochondria to accumulate a large amount of Ca^{2+}.

The accumulated Ca^{2+} in the matrix needs to be extruded to maintain a balance in the mitochondria. It is believed that exchangers are facilitating this function either by Na^+/Ca^{2+} exchange or by H^+/Ca^{2+} exchange. The Na^+/Ca^{2+} exchanger NCLX has been identified as a possible candidate in the mammalian system (De Marchi et al. 2014). The Arabidopsis cation/Ca^{2+} exchangers (CCX) are thought to be the plant counterparts but their primary subcellular localization outside the mitochondria puts this hypothesis in the debatable category (Wagner et al. 2016). The LETM1 protein falls in the category of H^+/Ca^{2+} exchanger (Jiang et al. 2009). Arabidopsis has two LETM1 homologues (LETM1 and LETM2) residing in the IMM (Zhang et al. 2012). Besides these proteins, the controversial permeability transition pore (PTP) can also release Ca^{2+} from mitochondria (Kanwar et al. 2020). However, transport through PTP may occur under drastic conditions.

Nucleus

Like the mitochondria, the nucleus is also a membrane-bound structure -a nuclear envelope, which consists of an outer nuclear membrane (ONM) and an inner nuclear membrane (INM) (Charpentier 2018). The nucleus is surrounded by the endoplasmic reticulum (ER), which can act as a Ca^{2+} supplier for the nucleus. Mostly the large aqueous channels known as nucleopore complexes (NPCs) serve as a connection between the nucleus and cytoplasm and hence can be the main entry point for Ca^{2+} into the nucleus (Charpentier 2018). Charpentier and colleagues have put forward a three-point hypothesis for nuclear Ca^{2+} import (Charpentier 2018). In case one, NPC simply import Ca^{2+} inside the nucleus to generate Ca^{2+} signal. In case two, the mechanism would involve activation of Ca^{2+} channels in the ONM and/or INM and then directing Ca^{2+} present in the nuclear envelope lumen into the nucleoplasm via NPC. And in case 3, the cytoplasmic Ca^{2+} signal would activate the Ca^{2+} channels in the nucleus to directly (or indirectly) by the action of a secondary component (Charpentier 2018).

The best case study to understand the nuclear Ca^{2+} signal is the *Medicago truncatula* system where the entire physiological process has been thoroughly worked out during root symbiosis. It starts with the perception of Nod factors (NFs) at the plasma membrane by lysine motif receptor-like kinases including Nod Factor Receptor 1 (NFR1) and NFR5 (Broghammer et al. 2012). These two then interact with receptor-like kinase Does not Make Infection 2 (DMI2) and DMI2 interacts with 3-hydroxy-3-methyl-glutaryl-coenzyme A reductase (HGMR1) to produce mevalonate, which can initiate nuclear Ca^{2+} signature (Charpentier 2018). It is speculated that the mevalonate and G-protein signaling pathway can further stimulate the production of another second messenger (it is still unknown and speculated to be cyclic nucleotide). This second messenger activates cyclic nucleotide-gated channel 15 (CNGC15)– and a potassium permeable channel Does not Make Infections (DMI1) complex, leading to Ca^{2+} release in the nucleus (Charpentier et al. 2016). A SERCA-type calcium ATPase (MCA8) act as the export pump to push the Ca^{2+} back into the lumen of the nuclear envelope (Charpentier et al. 2016). This particular Ca^{2+} signature generated inside the nucleus is decoded by calcium and calmodulin-dependent kinase (CCaMK) and it activates a downstream transcription factor CYCLOPS (Singh et al. 2014). Besides the *Medicago* system, work on *Lotus japonicas* has also revealed evidence for nuclear Ca^{2+} signaling. The nuclear envelope localized ion channels Castor and Pollux are involved in nuclear Ca^{2+} signature that is generated during root symbiosis with arbuscular mycorrhizal fungi and rhizobial bacteria (Charpentier et al. 2008). Figure 3.1 summarizes the players involved in the generation of organeller Ca^{2+} signature.

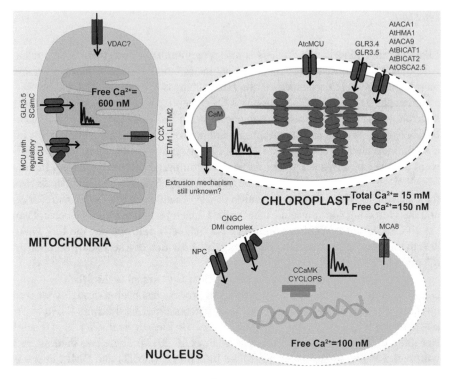

Fig. 3.1 Players involved in generating Ca²⁺ signature in the organelles of the cell. The chloroplast has AtcMCU, GLR3.4 and GLR3.5 as primary Ca²⁺ importer. Besides there may be other players involved in Arabidopsis are ACA1, ACA9, HMA1, BICAT1, BICAT2 and OSCA2.5. Extrusion mechanism is still unknown. CaM plays central role in transducing chloroplast Ca²⁺ signature. In the mitochondria VDAC may transport Ca²⁺ inside the inter-mitochondrial space. From there the MCU-MICU complex transport it into the matrix. GLR3.5 and SCamC may also perform similar function. Extrusion can be mediated by CCX and LETM exchangers. The nucleus probably uses NPC complexes for driving in Ca²⁺. The CNGC and DMI channels have a definitive role in Ca²⁺ uptake and MCA8 takes part in extrusion. The Ca²⁺ signature is decoded by CCaMK, which regulates the CYCLOPS transcription factor

Peroxisome

The stimuli that elicit cytosolic Ca²⁺ signature can also mediate peroxisomal Ca²⁺ accumulation (Costa et al. 2018). It has been shown that the enhanced Ca²⁺ in the peroxisomes controls the catalase 3 (CAT3) activity which in turn scavenges peroxisomal H_2O_2. The peroxisomal catalase is controlled by a calmodulin (CaM)-dependent mechanism (Yang and Poovaiah 2002). Besides the role in scavenging, CaM dependent mechanism is also functional in protein import and normal functionality of antioxidant and photorespiration enzymes present in the peroxisome (Costa et al. 2018). Moreover with identification of obvious physiological roles of Ca²⁺ signature in the peroxisome, the molecular identity of Ca²⁺ import/export candidate is still unknown.

Crosstalk of Ca^{2+} Signaling with Other Signaling Pathways

Abscisic acid (ABA) is a major stress hormone in plants. Recent research has shed light that Ca^{2+} signaling can crosstalk with ABA signaling to control plant responses (Edel and Kudla 2016). The common nodes where both the signaling pathways converge are- transporters and channels (AKT1, AKT2, NPF6.3, SLAC1/SLAH3), superoxide generators (RBOHD and RBOHF), and transcription factors (ABI5 and ABF1/4) (Edel and Kudla 2016). Besides, the C2-domain ABA-related (CAR) proteins (with functional Ca^{2+} binding C2 domain) can mediate recruitment of ABA receptors to the plasma membrane (Rodriguez et al. 2014; Diaz et al. 2016). The guard cell serves as the best model to study the integration of ABA and Ca^{2+} signaling. The basic model explains that ABA in the guard cell can trigger a Ca^{2+} signature that regulates various ion channels of the guard cell to show a synergistic effect of fast guard cell closing. Additionally many of the Ca^{2+} responsive kinases (CDPKs and CBL-CIPK module, more on them in the next chapter) have a direct role in either positive or negative regulation of ABA signaling by modulating downstream targets (Sanyal et al. 2019).

In addition to the core ABA signaling, Ca^{2+} play a critical role in maintaining nutrient homeostasis. The guard cell itself is a good example where K$^+$ channels are regulated to maintain the guard cell architecture (Sanyal et al. 2019). Outside the guard cell, the CBL-CIPK module is well known for its role in K$^+$ uptake by regulating AKT1, AKT2, HAK5 and TPK channels (Sanyal et al. 2019; Tang et al. 2020). Recent reports also suggest that the CBL-CIPK module plays a crucial role in regulating iron uptake (Tian et al. 2016). The CIPK11 mediated phosphor-regulation of the iron uptake master regulator (a transcription factor FIT) has opened up the further possibility that it might be a node for crosstalk between ABA, Ca^{2+} and nutrient signaling (Gratz et al. 2019). The nitrate uptake mechanism is also governed by the CBL-CIPK module by regulating the dual-affinity nitrate transporter (CHL1/ NRT1.1/NPF6.3) and ammonium uptake by regulating ammonium transporter (AMT1;2) (Ho et al. 2009; Straub et al. 2017). Finally, magnesium uptake is regulated by a complex interaction and crosstalk between the CBL-CIPK module and the SnRK2 kinases (Mogami et al. 2015; Tang et al. 2015).

In the biotic stress front too, we see active participation of Ca^{2+} signaling components. CDPKs play an important role in regulating plants response to biotic stress. The reactive oxygen species (ROS) are usually generated in the cell in response to pathogen attack. The NADPH oxidase/ respiratory burst homologs (RBOHs), play important role in ROS generation and are regulated by both CDPKs and CBL-CIPK module (Sanyal et al. 2019). Besides ROS can regulate Ca^{2+} signature and vice versa (Gilroy et al. 2014). Ca^{2+} can regulate a plethora of activities during the development of the pollen tube (Steinhorst and Kudla 2013). The steep gradient of Ca^{2+} that exists in the pollen tube help in its elongation and the rate at which it elongates. The stretch-activated Ca^{2+} channels (SAC), CNGCs, and GLRs all help in Ca^{2+} import into the cytoplasm of pollen tubes and are probable targets of CaBPs. We have already talked about the involvement of Ca^{2+} signaling in the nodulation pathway (in

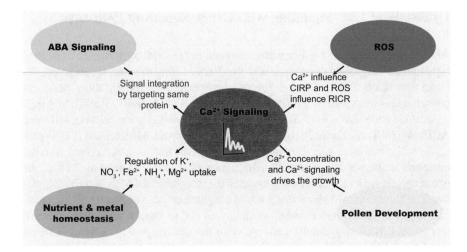

Fig. 3.2 Integration of Ca²⁺ signaling with other pathways in plants. The Ca^{2+} signaling and ABA signaling pathways integrate by targeting same targets there by increasing the complexity in modulation of these targets. Plant mineral/metal homeostasis is regulated at various levels by modulation of nutrient transport elements by Ca^{2+}-binding proteins. The ROS can influence ROS-induced calcium release (RICR) to modulate Ca^{2+} signature and Ca^{2+} can influence calcium-induced ROS production (CIRP). Pollen development is controlled largely by difference in Ca^{2+} gradient in the pollen and Ca^{2+} signaling events

the nuclear Ca^{2+} signaling segment) (Sanyal et al. 2019). Similarly, Ca^{2+} signaling is involved in the molecular processes that govern the interaction between plants and arbuscular-mycorrhizal (AM) fungi (Sanyal et al. 2019). Figure 3.2 describes the crosstalk of Ca^{2+} signaling with other pathways.

Conclusion

The Ca^{2+} signature hypothesis is an evolving concept. It started with the knowledge that stimulus generates transient Ca^{2+} increase in the cytoplasm and has gradually evolved to a point where we definitively know that these signatures are separately generated for the subcellular organelles. We also know that the Ca^{2+} signaling crosstalk with other pathways in the plant. From here we will move on to the next phase where information will be clear for the coordination of these responses.

References

C. Behrens, C. Blume, M. Senkler, H. Eubel, C. Peterhansel, H.P. Braun, The 'protein complex proteome' of chloroplasts in Arabidopsis thaliana. J. Proteome **91**, 73–83 (2013)

A. Broghammer, L. Krusell, M. Blaise, J. Sauer, J.T. Sullivan, N. Maolanon, M. Vinther, A. Lorentzen, E.B. Madsen, K.J. Jensen, P. Roepstorff, S. Thirup, C.W. Ronson, M.B. Thygesen, J. Stougaard, Legume receptors perceive the rhizobial lipochitin oligosaccharide signal molecules by direct binding. Proc. Natl. Acad. Sci. U. S. A. **109**, 13859–13864 (2012)

M. Charpentier, Calcium signals in the plant nucleus: Origin and function. J. Exp. Bot. **69**, 4165–4173 (2018)

M. Charpentier, R. Bredemeier, G. Wanner, N. Takeda, E. Schleiff, M. Parniske, Lotus japonicus CASTOR and POLLUX are ion channels essential for perinuclear calcium spiking in legume root endosymbiosis. Plant Cell **20**, 3467–3479 (2008)

M. Charpentier, J. Sun, T. Vaz Martins, G.V. Radhakrishnan, K. Findlay, E. Soumpourou, J. Thouin, A.A. Very, D. Sanders, R.J. Morris, G.E. Oldroyd, Nuclear-localized cyclic nucleotide-gated channels mediate symbiotic calcium oscillations. Science **352**, 1102–1105 (2016)

F. Chigri, J. Soll, U.C. Vothknecht, Calcium regulation of chloroplast protein import. Plant J. **42**, 821–831 (2005)

F. Chigri, F. Hormann, A. Stamp, D.K. Stammers, B. Bolter, J. Soll, U.C. Vothknecht, Calcium regulation of chloroplast protein translocation is mediated by calmodulin binding to Tic32. Proc. Natl. Acad. Sci. U. S. A. **103**, 16051–16056 (2006)

A. Costa, L. Navazio, I. Szabo, The contribution of organelles to plant intracellular calcium signalling. J. Exp. Bot **69**, 4175–4193 (2018)

E. Cutolo, N. Parvin, H. Ruge, N. Pirayesh, V. Roustan, W. Weckwerth, M. Teige, M. Grieco, V. Larosa, U.C. Vothknecht, The high light response in Arabidopsis requires the calcium sensor protein CAS, a target of STN7- and STN8-mediated phosphorylation. Front. Plant Sci. **10**, 974 (2019)

C. Dammann, A. Ichida, B. Hong, S.M. Romanowsky, E.M. Hrabak, A.C. Harmon, B.G. Pickard, J.F. Harper, Subcellular targeting of nine calcium-dependent protein kinase isoforms from Arabidopsis. Plant Physiol. **132**, 1840–1848 (2003)

U. De Marchi, J. Santo-Domingo, C. Castelbou, I. Sekler, A. Wiederkehr, N. Demaurex, NCLX protein, but not LETM1, mediates mitochondrial Ca2+ extrusion, thereby limiting Ca2+−induced NAD(P)H production and modulating matrix redox state. J. Biol. Chem. **289**, 20377–20385 (2014)

D. De Stefani, A. Raffaello, E. Teardo, I. Szabo, R. Rizzuto, A forty-kilodalton protein of the inner membrane is the mitochondrial calcium uniporter. Nature **476**, 336–340 (2011)

M. Diaz, M.J. Sanchez-Barrena, J.M. Gonzalez-Rubio, L. Rodriguez, D. Fernandez, R. Antoni, C. Yunta, B. Belda-Palazon, M. Gonzalez-Guzman, M. Peirats-Llobet, M. Menendez, J. Boskovic, J.A. Marquez, P.L. Rodriguez, A. Albert, Calcium-dependent oligomerization of CAR proteins at cell membrane modulates ABA signaling. Proc. Natl. Acad. Sci. U. S. A. **113**, E396–E405 (2016)

A.N. Dodd, J. Kudla, D. Sanders, The language of calcium signaling. Annu. Rev. Plant Biol. **61**, 593–620 (2010)

T.P.J. Dunkley, S. Hester, I.P. Shadforth, J. Runions, T. Weimar, S.L. Hanton, J.L. Griffin, C. Bessant, F. Brandizzi, C. Hawes, R.B. Watson, P. Dupree, K.S. Lilley, Mapping the Arabidopsis organelle proteome. Proc. Natl. Acad. Sci. U. S. A. **103**, 6518–6523 (2006)

K.H. Edel, J. Kudla, Integration of calcium and ABA signaling. Curr. Opin. Plant Biol. **33**, 83–91 (2016)

J. Frank, R. Happeck, B. Meier, M.T.T. Hoang, J. Stribny, G. Hause, H. Ding, P. Morsomme, S. Baginsky, E. Peiter, Chloroplast-localized BICAT proteins shape stromal calcium signals and are required for efficient photosynthesis. New Phytol. **221**, 866–880 (2019)

S. Gilroy, N. Suzuki, G. Miller, W.G. Choi, M. Toyota, A.R. Devireddy, R. Mittler, A tidal wave of signals: Calcium and ROS at the forefront of rapid systemic signaling. Trends Plant Sci. **19**, 623–630 (2014)

R. Gratz, P. Manishankar, R. Ivanov, P. Koster, I. Mohr, K. Trofimov, L. Steinhorst, J. Meiser, H.J. Mai, M. Drerup, S. Arendt, M. Holtkamp, U. Karst, J. Kudla, P. Bauer, T. Brumbarova, CIPK11-dependent phosphorylation modulates FIT activity to promote Arabidopsis iron acquisition in response to calcium signaling. Dev. Cell **48**, 726–740 (2019)

S. Helm, D. Dobritzsch, A. Rodiger, B. Agne, S. Baginsky, Protein identification and quantification by data-independent acquisition and multi-parallel collision-induced dissociation mass spectrometry (MS(E)) in the chloroplast stroma proteome. J. Proteome **98**, 79–89 (2014)

C.H. Ho, S.H. Lin, H.C. Hu, Y.F. Tsay, CHL1 functions as a nitrate sensor in plants. Cell **138**, 1184–1194 (2009)

L. Huang, T. Berkelman, A.E. Franklin, N.E. Hoffman, Characterization of a gene encoding a Ca(2+)-ATPase-like protein in the plastid envelope. Proc. Natl. Acad. Sci. U. S. A. **90**, 10066–10070 (1993)

H.W. Jarrett, C.J. Brown, C.C. Black, M.J. Cormier, Evidence that calmodulin is in the chloroplast of peas and serves a regulatory role in photosynthesis. J. Biol. Chem. **257**, 13795–13804 (1982)

R. Ji, L. Zhou, J. Liu, Y. Wang, L. Yang, Q. Zheng, C. Zhang, B. Zhang, H. Ge, Y. Yang, F. Zhao, S. Luan, W. Lan, Calcium-dependent protein kinase CPK31 interacts with arsenic transporter AtNIP1;1 and regulates arsenite uptake in Arabidopsis thaliana. PLoS One **12**, e0173681 (2017)

D. Jiang, L. Zhao, D.E. Clapham, Genome-wide RNAi screen identifies Letm1 as a mitochondrial Ca2+/H+ antiporter. Science **326**, 144–147 (2009)

P. Kanwar, H. Samtani, S.K. Sanyal, A.K. Srivastava, P. Suprasanna, G.K. Pandey, VDAC and its interacting partners in plant and animal systems: An overview. Crit. Rev. Biotechnol. **40**(5), 1–18 (2020)

M.C. Kim, W.S. Chung, D.J. Yun, M.J. Cho, Calcium and calmodulin-mediated regulation of gene expression in plants. Mol. Plant **2**, 13–21 (2009)

G. Kreimer, M. Melkonian, J.A. Holtum, E. Latzko, Characterization of calcium fluxes across the envelope of intact spinach chloroplasts. Planta **166**, 515–523 (1985)

J. Kudla, O. Batistic, K. Hashimoto, Calcium signals: The lead currency of plant information processing. Plant Cell **22**, 541–563 (2010)

J. Lee, H. Lee, J. Kim, S. Lee, D.H. Kim, S. Kim, I. Hwang, Both the hydrophobicity and a positively charged region flanking the C-terminal region of the transmembrane domain of signal-anchored proteins play critical roles in determining their targeting specificity to the endoplasmic reticulum or endosymbiotic organelles in Arabidopsis cells. Plant Cell **23**, 1588–1607 (2011)

Z.Y. Li, Z.S. Xu, G.Y. He, G.X. Yang, M. Chen, L.C. Li, Y. Ma, The voltage-dependent anion channel 1 (AtVDAC1) negatively regulates plant cold responses during germination and seedling development in Arabidopsis and interacts with calcium sensor CBL1. Int. J. Mol. Sci. **14**, 701–713 (2013)

S. Malmstrom, P. Askerlund, M.G. Palmgren, A calmodulin-stimulated Ca2+-ATPase from plant vacuolar membranes with a putative regulatory domain at its N-terminus. FEBS Lett. **400**, 324–328 (1997)

S. Masuda, K. Mizusawa, T. Narisawa, Y. Tozawa, H. Ohta, K. Takamiya, The bacterial stringent response, conserved in chloroplasts, controls plant fertilization. Plant Cell Physiol. **49**, 135–141 (2008)

M.R. McAinsh, A.M. Hetherington, Encoding specificity in Ca^{2+} signalling systems. Trends Plant Sci. **3**, 32–36 (1998)

M.R. McAinsh, C. Brownlee, A.M. Hetherington, Calcium ions as second messengers in guard cell signal transduction. Physiol. Plant. **100**, 16–29 (1997)

A.J. Miller, D. Sanders, Depletion of cytosolic free calcium induced by photosynthesis. Nature **326**, 397–400 (1987)

J. Mogami, Y. Fujita, T. Yoshida, Y. Tsukiori, H. Nakagami, Y. Nomura, T. Fujiwara, S. Nishida, S. Yanagisawa, T. Ishida, F. Takahashi, K. Morimoto, S. Kidokoro, J. Mizoi, K. Shinozaki, K. Yamaguchi-Shinozaki, Two distinct families of protein kinases are required for plant growth under high external Mg2+ concentrations in Arabidopsis. Plant Physiol. **167**, 1039–1057 (2015)

I. Moreno, L. Norambuena, D. Maturana, M. Toro, C. Vergara, A. Orellana, A. Zurita-Silva, V.R. Ordenes, AtHMA1 is a thapsigargin-sensitive Ca2+/heavy metal pump. J. Biol. Chem. **283**, 9633–9641 (2008)

L. Navazio, E. Formentin, L. Cendron, I. Szabò, Chloroplast calcium signaling in the spotlight. Front. Plant Sci. **11**, 186 (2020)

G. Pedriali, A. Rimessi, L. Sbano, C. Giorgi, M.R. Wieckowski, M. Previati, P. Pinton, Regulation of endoplasmic reticulum–mitochondria Ca2+ transfer and its importance for anti-cancer therapies. Front. Oncol. **7**, 180 (2017)

M. Plovanich, R.L. Bogorad, Y. Sancak, K.J. Kamer, L. Strittmatter, A.A. Li, H.S. Girgis, S. Kuchimanchi, J. De Groot, L. Speciner, N. Taneja, J. Oshea, V. Koteliansky, V.K. Mootha, MICU2, a paralog of MICU1, resides within the mitochondrial uniporter complex to regulate calcium handling. PLoS One **8**, e55785 (2013)

L. Rodriguez, M. Gonzalez-Guzman, M. Diaz, A. Rodrigues, A.C. Izquierdo-Garcia, M. Peirats-Llobet, M.A. Fernandez, R. Antoni, D. Fernandez, J.A. Marquez, J.M. Mulet, A. Albert, P.L. Rodriguez, C2-domain abscisic acid-related proteins mediate the interaction of PYR/PYL/RCAR abscisic acid receptors with the plasma membrane and regulate abscisic acid sensitivity in Arabidopsis. Plant Cell **26**, 4802–4820 (2014)

M.H. Roh, R. Shingles, M.J. Cleveland, R.E. McCarty, Direct measurement of clcium transport across chloroplast inner-envelope Vesicles1. Plant Physiol. **118**, 1447–1454 (1998)

Y. Sancak, A.L. Markhard, T. Kitami, E. Kovacs-Bogdan, K.J. Kamer, N.D. Udeshi, S.A. Carr, D. Chaudhuri, D.E. Clapham, A.A. Li, S.E. Calvo, O. Goldberger, V.K. Mootha, EMRE is an essential component of the mitochondrial calcium uniporter complex. Science **342**, 1379–1382 (2013)

S.K. Sanyal, S. Mahiwal, G.K. Pandey, Calcium Signaling: A Communication Network that Regulates Cellular Processes, in *Sensory Biology of Plants*, ed. by S. Sopory, (Springer, Singapore, 2019), pp. 279–309

M. Schiott, S.M. Romanowsky, L. Baekgaard, M.K. Jakobsen, M.G. Palmgren, J.F. Harper, A plant plasma membrane Ca2+ pump is required for normal pollen tube growth and fertilization. Proc. Natl. Acad. Sci. U. S. A. **101**, 9502–9507 (2004)

D. Seigneurin-Berny, A. Gravot, P. Auroy, C. Mazard, A. Kraut, G. Finazzi, D. Grunwald, F. Rappaport, A. Vavasseur, J. Joyard, P. Richaud, N. Rolland, HMA1, a new Cu-ATPase of the chloroplast envelope, is essential for growth under adverse light conditions. J. Biol. Chem. **281**, 2882–2892 (2006)

S. Singh, K. Katzer, J. Lambert, M. Cerri, M. Parniske, CYCLOPS, a DNA-binding transcriptional activator, orchestrates symbiotic root nodule development. Cell Host Microbe **15**, 139–152 (2014)

S. Stael, A.G. Rocha, A.J. Robinson, P. Kmiecik, U.C. Vothknecht, M. Teige, Arabidopsis calcium-binding mitochondrial carrier proteins as potential facilitators of mitochondrial ATP-import and plastid SAM-import. FEBS Lett. **585**, 3935–3940 (2011)

S. Stael, B. Wurzinger, A. Mair, N. Mehlmer, U.C. Vothknecht, M. Teige, Plant organellar calcium signalling: An emerging field. J. Exp. Bot. **63**, 1525–1542 (2012)

L. Steinhorst, J. Kudla, Calcium – A central regulator of pollen germination and tube growth. Biochim. Biophys. Acta **1833**, 1573–1581 (2013)

T. Straub, U. Ludewig, B. Neuhauser, The kinase CIPK23 inhibits ammonium transport in Arabidopsis thaliana. Plant Cell **29**, 409–422 (2017)

E. Sundberg, J.G. Slagter, I. Fridborg, S.P. Cleary, C. Robinson, G. Coupland, ALBINO3, an Arabidopsis nuclear gene essential for chloroplast differentiation, encodes a chloroplast protein that shows homology to proteins present in bacterial membranes and yeast mitochondria. Plant Cell **9**, 717–730 (1997)

H. Takahashi, A. Watanabe, A. Tanaka, S.N. Hashida, M. Kawai-Yamada, K. Sonoike, H. Uchimiya, Chloroplast NAD kinase is essential for energy transduction through the xanthophyll cycle in photosynthesis. Plant Cell Physiol. **47**, 1678–1682 (2006)

R.J. Tang, F.G. Zhao, V.J. Garcia, T.J. Kleist, L. Yang, H.X. Zhang, S. Luan, Tonoplast CBL-CIPK calcium signaling network regulates magnesium homeostasis in Arabidopsis. Proc. Natl. Acad. Sci. U. S. A. **112**, 3134–3139 (2015)

R.J. Tang, F.G. Zhao, Y. Yang, C. Wang, K. Li, T.J. Kleist, P.G. Lemaux, S. Luan, A calcium signalling network activates vacuolar K(+) remobilization to enable plant adaptation to low-K environments. Nat. Plants **6**, 384–393 (2020)

E. Teardo, A. Segalla, E. Formentin, M. Zanetti, O. Marin, G.M. Giacometti, F. Lo Schiavo, M. Zoratti, I. Szabo, Characterization of a plant glutamate receptor activity. Cell. Physiol. Biochem. **26**, 253–262 (2010)

E. Teardo, E. Formentin, A. Segalla, G.M. Giacometti, O. Marin, M. Zanetti, F. Lo Schiavo, M. Zoratti, I. Szabo, Dual localization of plant glutamate receptor AtGLR3.4 to plastids and plasmamembrane. Biochim. Biophys. Acta **1807**, 359–367 (2011)

E. Teardo, L. Carraretto, S. De Bortoli, A. Costa, S. Behera, R. Wagner, F. Lo Schiavo, E. Formentin, I. Szabo, Alternative splicing-mediated targeting of the Arabidopsis GLUTAMATE RECEPTOR3.5 to mitochondria affects organelle Morphology1. Plant Physiol. **167**, 216–227 (2015)

E. Teardo, L. Carraretto, R. Moscatiello, E. Cortese, M. Vicario, M. Festa, L. Maso, S. De Bortoli, T. Cali, U.C. Vothknecht, E. Formentin, L. Cendron, L. Navazio, I. Szabo, A chloroplast-localized mitochondrial calcium uniporter transduces osmotic stress in Arabidopsis. Nat. Plants **5**, 581–588 (2019)

Q. Tian, X. Zhang, A. Yang, T. Wang, W.H. Zhang, CIPK23 is involved in iron acquisition of Arabidopsis by affecting ferric chelate reductase activity. Plant Sci. **246**, 70–79 (2016)

S. Wagner, S. Behera, S. De Bortoli, D.C. Logan, P. Fuchs, L. Carraretto, E. Teardo, L. Cendron, T. Nietzel, M. Fussl, F.G. Doccula, L. Navazio, M.D. Fricker, O. Van Aken, I. Finkemeier, A.J. Meyer, I. Szabo, A. Costa, M. Schwarzlander, The EF-hand Ca2+ binding protein MICU choreographs mitochondrial Ca2+ dynamics in Arabidopsis. Plant Cell **27**, 3190–3212 (2015)

S. Wagner, S. De Bortoli, M. Schwarzlander, I. Szabo, Regulation of mitochondrial calcium in plants versus animals. J. Exp. Bot. **67**, 3809–3829 (2016)

M. Waldeck-Weiermair, R. Malli, W. Parichatikanond, B. Gottschalk, C.T. Madreiter-Sokolowski, C. Klec, R. Rost, W.F. Graier, Rearrangement of MICU1 multimers for activation of MCU is solely controlled by cytosolic Ca(2). Sci. Rep. **5**, 15602 (2015)

D. Wang, Y. Xu, Q. Li, X. Hao, K. Cui, F. Sun, Y. Zhu, Transgenic expression of a putative calcium transporter affects the time of Arabidopsis flowering. Plant J. **33**, 285–292 (2003)

S. Weinl, K. Held, K. Schlucking, L. Steinhorst, S. Kuhlgert, M. Hippler, J. Kudla, A plastid protein crucial for Ca2+−regulated stomatal responses. New Phytol. **179**, 675–686 (2008)

T. Yang, B.W. Poovaiah, Hydrogen peroxide homeostasis: Activation of plant catalase by calcium/calmodulin. Proc. Natl. Acad. Sci. U. S. A. **99**, 4097–4102 (2002)

B. Zhang, C. Carrie, A. Ivanova, R. Narsai, M.W. Murcha, O. Duncan, Y. Wang, S.R. Law, V. Albrecht, B. Pogson, E. Giraud, O. Van Aken, J. Whelan, LETM proteins play a role in the accumulation of mitochondrially encoded proteins in Arabidopsis thaliana and AtLETM2 displays parent of origin effects. J. Biol. Chem. **287**, 41757–41773 (2012)

Chapter 4
Calcium-Binding Proteins- "Decoders of Ca²⁺ Signature"

Contents

Introduction

We had a very brief glimpse of the CaBPs in the last chapter. Ca^{2+} signature that is generated in the cell carries a "coded message" that has to be delivered to the specific target to elicit a response. The cell has a whole repertoire of Ca^{2+} sensors, which can decode the message of a particular Ca^{2+} signature and transduce the signal to a particular target (Luan 2009; Dodd et al. 2010; Pandey et al. 2014). These proteins are collectively called CaBPs and their defining feature is the presence of EF-hand motifs, which usually vary in number (Gifford et al. 2007). The EF-hand motif was first discovered in parvalbumin and its crystal structure indicated that it was formed by E and F helices and so the name was derived (Kretsinger and Nockolds 1973). It is a helix-loop-helix structure (two α-helices bridged by a Ca^{2+}-chelation loop). This loop usually is rich in negatively charged amino acids (like Asp or Glu) that can coordinate oxygen atom (McPhalen et al. 1991). This helps the loop to coordinate and accommodate the Ca^{2+} ion. Ca^{2+} prefers a pentagonal bipyramidal arrangement of the amino acids (the coordinating ligands) within the loop of the EF-hand (Strynadka and James 1989). The canonical EF-hand loop comprises of nine residues and provides five Ca^{2+} coordinating ligands. The remaining two are provided by the side chain of an acidic amino acid (usually Glu) present at 12th position in the loop. Taken together this forms the core group of 12 amino acid residues that coordinate Ca^{2+} (Gifford et al. 2007). The ligands are notated based on their linear position in the loop and the tertiary geometry imposed by the alignment on the axes of the bipyramid, and hence 1(+X), 3(+Y), 5(+Z), 7(-Y), 9(−X), 12(−Z)

G. K. Pandey, S. K. Sanyal, *Functional Dissection of Calcium Homeostasis and Transport Machinery in Plants*, SpringerBriefs in Plant Science, https://doi.org/10.1007/978-3-030-58502-0_4

(Strynadka and James 1989). The ligands at position 1, 3, 5 and 12 co-ordinate Ca^{2+} through their side chain, the ligand at 7 uses its backbone and the ligand 9 uses side chain to bind water molecule, which coordinates the Ca^{2+} (Gifford et al. 2007). At position 4 and 6 conserved Gly is present for the conformational requirement of the loop (La Verde et al. 2018) (indicated in Fig. 4.1). However, there are several examples where EF-hand does not use the typical geometry and improvises to bind Ca^{2+} (Gifford et al. 2007).

The defining property of the EF-hand containing CaBPs (henceforth only CaBPs) is the conformational change after Ca^{2+} binding (Batistic et al. 2011). In absence of Ca^{2+}, the EF-hand usually adopt a closed conformation and Ca^{2+} binding opens up

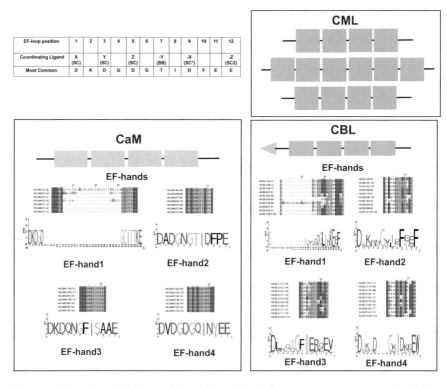

Fig. 4.1 EF-hand containing CaBPs of plants (the Ca²⁺ sensor group and the CBLs). The amino acids that most commonly appear in the EF-hand sequence with their positioning in the loop. The bracketed words indicate the side chain (SC), back bone (BB) and side chain (SC*) used for coordinating Ca^{2+} by the ligands at this particular position. The information is adopted from (Gifford et al. 2007). The CaM proteins are indicated with their four EF-hands and a multiple sequence alignment (MSA) of their EF-hands. The amino acids of the seven CaM EF-hands are invariant. Another plant Ca^{2+} sensor CBL are indicated with their N-terminal targeting sequence and EF-hands. The MSA here shows that there is a great degree of variation among the EF-hand sequences indicating that these proteins can function in different Ca^{2+} signatures. The CML proteins with their variable EF-hands are indicated. The sequences were taken from TAIR database and MSA performed by MEGA10 and viewed using Jalview. The consensus was viewed using WebLogo

the hand exposing the hydrophobic clefts that can bind targets (Sanyal et al. 2015). This EF-hand containing CaBPs of plants can be divided into two main groups: the sensor relays and sensor responders (Sanders et al. 2002; Hashimoto and Kudla 2011). Each class are discussed in the following section.

Calmodulins and Calmodulin-Like Protein Sensors

The sensor relays, as the name suggests, will only act to relay the message encoded in a Ca^{2+} signature. Usually, they do not possess any functional domains (other than EF-hands and hence possess no biochemical/enzymatic function) and mediate conformational change of the target (binding/interacting proteins) to modulate their activity (DeFalco et al. 2010). The Calmodulin (CaMs) and Calmodulin-like (CMLs) of plants belong to this group. The Arabidopsis genome has seven CaMs and about fifty CML genes and these genes have been reported in other plant species as well (Snedden and Fromm 2001; McCormack and Braam 2003; Bouche et al. 2005; Bender and Snedden 2013; Zeng et al. 2015).

The CaMs generally possess 4 EF-hands that are similar to the animal counterparts and resemble the canonical structure (indicated in Fig. 4.1). Ca^{2+} binding allows the exposure of hydrophobic clefts within the CaM proteins, which allow the downstream targets with CaM binding domains (CaMBDs) (usually 16–35 amino acid long) to interact with CaM (Ikura and Ames 2006). The CMLs in comparison have a variable number of EF-hands (ranging from 1 to 6) and also show a considerable deviation from the canonical sequence (Bender and Snedden 2013) (indicated in Fig. 4.1). The 12th amino acid of the CML is usually Asp (instead of canonical Gln), which may alter its binding preference to magnesium (Mg^{2+}) instead of Ca^{2+}. The CMLs usually have a myristoylation motif present at their N-terminus (not seen in CaMs) that help them in subcellular targeting (DeFalco et al. 2010). Only few CML's EF-hands have been proven to bind Ca^{2+} through experiments. Besides the difference in their EF-hand sequences, the CaMs and CMLs differ in their length, the CMLs being longer with an extension of both their N- and C- terminals (DeFalco et al. 2010).

A large number of targets interact with CaMs and CMLs (the list includes kinases, phosphatases, receptors, transport machinery components, metabolic enzymes and transcription factors) (DeFalco et al. 2010; Hashimoto and Kudla 2011; Zeng et al. 2015). Table 4.1 gives a brief description of CaM targets. We have already talked about CaM and NADK interaction in the previous chapter. Besides these the CaM/CML has been implicated in development, hormone signaling abiotic stress (heat stress, cold stress, salt and drought stress, heavy metal stress) and biotic stress (Zeng et al. 2015). Targets for CML are described in Table 4.2. There is also definitive proof that they crosstalk with ROS signaling components to mediate plant stress signaling (indicated in Fig. 4.3).

Table 4.1 The table describes the targets regulated by CaM. It includes all classes of protein targets of CaM

Protein Classification	Target	Physiological role
Phosphatases	AtPP7	Heat
Kinases	AtCBK3/CRK1, AtCRCK1, AtCRLK1	Heat Cold, Salt, ABA, H_2O_2
Transcription factor and co-factors	AtSR1/CAMTA3, AtSR2/CAMTA1, AtSR2/ CAMTA2, AtMYB2, AtABF2/AREB1, AtCBP60g, AtGTL1, AtGT2L, AtBT1–5, AtBT2	Heat Cold, salt, drought, Dehydration, ABA, H_2O_2, SA
Ion transporter	AtCNGC1 AtNHX1 AtACA4	Heavy metal, Salt
Metabolic enzyme	AtCAT3	Oxidative stress

Table data from (Zeng et al. 2015). Detailed references for each target in (Zeng et al. 2015)

Table 4.2 A table describing the targets regulated by CML

CML	Target	Physiological Role
CML8	BRI1, ZAR1, IQD1, PEN3	Plant immunity
CML9	PPR2, IQD1, PEN3, ILK1	Signaling hub
CML10	PM-MUTASE	Abiotic stress (Oxidative stress)
CML12	PINOID, PEN3	Development; Plant immunity
CML18	NHX1, CBP60C	Abiotic stress (Salt)
CML19	RAD4, SAC3b, DSS1	Abiotic stress (UV-damage)
CML20	TON1, SAC3, UCH	Abiotic stress (Drought Stress)
CML24	ATG4b	Signaling hub
CML37	PEN3	Signaling hub
CML38	RALF1, PEN3	Signaling hub
CML42	KIC	Signaling hub

Table data and signaling hub classification according to (La Verde et al. 2018). Detailed references for each target in (La Verde et al. 2018)

Calcineurin B-Like Protein and CBL-Interacting Protein Kinase Module

The next entry in the CaBP/Ca^{2+} sensors repertoire is not a single protein but a module comprising of a Ca^{2+} sensor Calcineurin B-like (CBL) and CBL-interacting protein kinase (CIPK). Its classification should have been in the sensor relay group, as the core CBL does not have any special functional domain like the CaMs, but its association with CIPKs, which can modulate targets argues in the favor of placing it in the sensor responder group. The module is seen right from algal species to embryophytes with the evolutionarily advanced species having more number of CBLs and CIPKs (usually CIPKs are more in number compare to CBLs) (Kleist et al. 2014; Pandey et al. 2014; Beckmann et al. 2016; Edel et al. 2017). The CBLs and CIPKs can have multiple interactions between themselves to possibly give rise to multiple modules. And this is a special feature of the module that increases its versatility (Pandey 2008; Pandey et al. 2014; Sanyal et al. 2015, 2016).

The CBLs, as already mentioned are Ca^{2+} sensors, and so have four EF-hands and an N-terminal localized subcellular targeting motif (Batistic et al. 2010) (indicated in Fig. 4.1). A novel phosphorylation motif "PFPF" was identified later in them which is phosphorylated by CIPK to enhance the interaction between CBL and CIPK and further activate the kinase activity of the CIPK (Du et al. 2011; Hashimoto et al. 2012). The localization motifs are of three types-Type I (allowing for plasma membrane localization), Type II (allowing tonoplast localization) and Type III (allowing dual localization). The Type I seems to be the most ancient motif and the other two have evolved later (Kleist et al. 2014). These motifs guide the localization of the module inside the cell, although more investigation is necessary to explain how downstream targets outside the range of these motifs are modulated. The EF-hands of CBLs is non-canonical, with EF-hand 1 having 14 amino acids instead of 12. Although there are shreds of evidences that these proteins can bind to Ca^{2+}, but the more experimental proof is required to explain the *in vivo* Ca^{2+} binding (Sanchez-Barrena et al. 2013). CIPKs have the normal architecture of Ser/Thr kinases with the ATP binding pocket and activation loop. Within the loop, there are three specific amino acids (Ser, Thr, Tyr) that are thought to be phosphorylated (by some other kinase) to activate the kinase further (Gong et al. 2002; Sanyal et al. 2015, 2020; Yadav et al. 2018). The unique identity of a CIPK is given by the NAF domain in its regulatory domain. This domain helps the CIPK to bind to a CBL (Albrecht et al. 2001) (indicated in Fig. 4.2). The Ca^{2+} binding to CBL allows the NAF dependent binding of CIPK and the resultant conformational change removes the regulatory domain (of CIPK) from the kinase domain (of CIPK) to make it an active kinase capable of target regulation (Sanyal et al. 2020). The NAF domain is closely followed by PPI domain, which is the target site for phosphatases (mostly clade 2A PP2Cs and a clade B PP2C) (Ohta et al. 2003; Singh et al. 2018). Most of these phosphatases have been found to negatively regulate the action of CBL-CIPK module (reviewed in (Sanyal et al. 2020)).

The module usually employs phosphoregulation of target proteins (Sanyal et al. 2020). Once it reaches the target site to transduce the message encoded in a Ca^{2+} signature, it would usually phosphorylate residue(s) in the target that is important for the regulation. There are a few examples of regulation without phosphorylation, but generally, these reports are an aberration (Sanyal et al. 2015). There are plethora of pathways CBL-CIPK module regulates. They are involved in ABA signaling and abiotic stress, nutrition uptake and metal homeostasis, ROS regulation and biotic stress and plant development (Pandey 2008; Pandey et al. 2014, 2015; Sanyal et al. 2019) (indicated in Fig. 4.3).

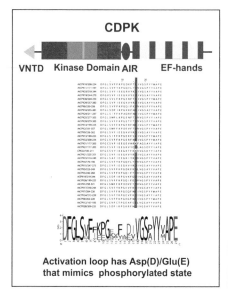

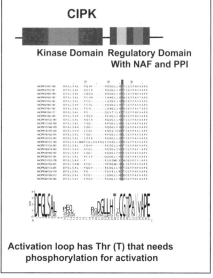

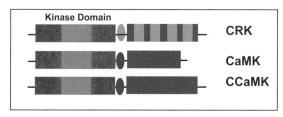

Fig. 4.2 The sensor responder, CIPKs and other Ca²⁺ regulated kinases of plants. The CDPKs are represented with their VNTD, kinase domain, AIR and CLD. The activation loop of the CDPK has a substitution of either Asp (D)/Glu (E) in a very important position which is indicated through a blue mark in the MSA. The CIPKs are indicated with their kinase domain and regulatory domain which is different from CDPK and has a NAF and PPI motif. The activation loop is also different from CDPK and contains Thr (T) that needs phosphorylation for activation. The other kinases-CRK (with degenerate EF-hands and no junction domain), CaMK with no EF-hands and CCaMK with visinin-like domains are represented. The sequences were taken from TAIR database and MSA performed by MEGA10 and viewed using Jalview. The consensus was viewed using WebLogo

Ca²⁺ Dependent Protein Kinase

The Ca²⁺ dependent protein kinases (CDPK, aka CPK in Arabidopsis) are the members of sensor responder group of plant CaBPs (Hrabak et al. 2003; Takahashi and Ito 2011; Bender et al. 2018). Evolutionarily, they are considered to be a fusion product of CaM and a CaM-dependent protein kinase (CaMK) (Harper et al. 1991). Like the CBL-CIPK module, the CDPKs also show a similar trend in their increased number from algae to embryophytes (Hamel et al. 2014) .

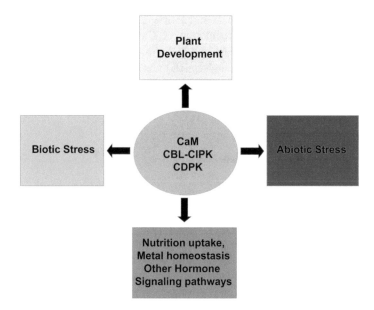

Fig. 4.3 Biological role played by CaBPs in plants. Research so far has indicated that the CaBPs (mainly the CaMs, CDPKs and CBL-CIPK module) can transduce information carried by plethora of Ca²⁺ signatures. The pathways they cater are abiotic and biotic stress, plant development, hormone signaling and metal homeostasis. Information is not complete for some CaBPs (like CMLs), but they too take part in important pathways necessary for plant survival

The CDPKs, like the CIPKs, are Ser/Thr kinases. The protein consists of a variable N-terminal domain (VNTD), followed by a kinase domain, then the autoinhibitory region (AIR aka pseudo-substrate segment), which connect the kinase domain with the CaM-like domain (CLD) (Takahashi and Ito 2011; Hamel et al. 2014) (indicated in Fig. 4.2). The VNTD as the name suggests is variable in its length and contains the localization motif that guides the subcellular targeting of a CDPK (Takahashi and Ito 2011). Besides this core function, there are proofs that the VNTD can perform some additional function like substrate binding and phosphorylation of substrate (Ito et al. 2010; Asai et al. 2013; Ying et al. 2017). The kinase domain of CDPK contains the activation loop (like any Ser/Thr kinase), but unlike the CIPKs, the CDPKs have acidic residue(s) in the loop (in place of the phosphorylation specific residue of other kinases) that mimic already phosphorylated state (Takahashi and Ito 2011; Ito et al. 2017). The AIR domain inhibits the CDPK activity by binding to the active site. The CLD domain comes next which usually has four EF-hands and is responsible for the Ca²⁺ binding by the CDPKs. The AIR and CLD together form the CDPK activation domain (CAD) and Ca²⁺ binding removes the CAD domain from the kinase domain removing the autoinhibition (Takahashi and Ito 2011; Hamel et al. 2014; Bender et al. 2018).

The CDPKs also regulate their targets by phosphorylation. They are actively involved in biotic stress, abiotic stress and hormonal signaling (Romeis et al. 2001;

Boudsocq and Sheen 2013; Romeis and Herde 2014). There is also evidence for involvement of CDPKs in the regulation of Ca^{2+} homeostasis and metabolism (Simeunovic et al. 2016). The guard cell is a good example where we can see that integration of ABA signaling (through SnRK) and Ca^{2+} signaling (CDPK) that help in fast closing of the guard cell in response to stress stimuli (Huang et al. 2019) (indicated in Fig. 4.3).

Other Ca^{2+} Dependent Kinases

In addition to the important CaBPs discussed above, some other candidates play an important role in Ca^{2+} signaling. There is a whole set of kinases that either directly regulated by Ca^{2+} or by CaM. The CDPK related kinases (CRK) have a similar sequence like the CDPKs but they possess degenerated EF-hands (Chae et al. 2010). They probably do not have an autoinhibitory domain and may be activated without Ca^{2+} (Das and Pandey, 2010). However, there is evidence of Ca^{2+}/CaM dependent activation (Sanyal et al. 2015). The CaM-dependent protein kinases (CaMKs) have kinase domain and auto-inhibitory but do not possess any EF-hands (Hrabak et al. 2003). They are regulated by the CaMs and are more prevalent in animals (they do not have any presence in Arabidopsis). The Ca^{2+} or Ca^{2+}/CaM regulated kinases (CCaMK) have visinin-like domains in their C-terminal in place of EF-hands to bind Ca^{2+}. Again like the CaMKs, the CCaMKs are absent in Arabidopsis but are an important component of nuclear Ca^{2+} signaling as reported in *Lotus japonicas* (Hrabak et al. 2003; Chae et al. 2010; Charpentier 2018). Figure 4.2 depicts the basic architecture of this class of kinases. Besides these core proteins several other classes of CaBPs in plants exist, some of them are discussed in the previous chapter which significantly enhances the plant Ca^{2+} sensor repertoire.

Conclusion

The major Ca^{2+} signature decoding work in plants is performed by the CaM, CBL-CIPK module and CDPKs. Each of them has their specific optimized functioning and increase the complexity of decoding of Ca^{2+} signaling. The current data does not fully explain the Ca^{2+} binding efficiency of the EF-hands for the majority of the CaBPs. However, it is undeniable that these CaBPs regulate important physiological processes in plants and hence are important part of plant Ca^{2+} signaling toolkit.

References

V. Albrecht, O. Ritz, S. Linder, K. Harter, J. Kudla, The NAF domain defines a novel protein-protein interaction module conserved in Ca^{2+} regulated kinases. EMBO J. **20**, 1051–1063 (2001)

S. Asai, T. Ichikawa, H. Nomura, M. Kobayashi, Y. Kamiyoshihara, H. Mori, Y. Kadota, C. Zipfel, J.D. Jones, H. Yoshioka, The variable domain of a plant calcium-dependent protein kinase (CDPK) confers subcellular localization and substrate recognition for NADPH oxidase. J. Biol. Chem. **288**, 14332–14340 (2013)

O. Batistic, R. Waadt, L. Steinhorst, K. Held, J. Kudla, CBL-mediated targeting of CIPKs facilitates the decoding of calcium signals emanating from distinct cellular stores. Plant J. **61**, 211–222 (2010)

O. Batistic, K.N. Kim, T. Kleist, J. Kudla, S. Luan, The CBL–CIPK Network for Decoding Calcium Signals in Plants, in *Coding and Decoding of Calcium Signals in Plants*, ed. by S. Luan, (Springer, Berlin, 2011), pp. 235–258

L. Beckmann, K.H. Edel, O. Batistic, J. Kudla, A calcium sensor – protein kinase signaling module diversified in plants and is retained in all lineages of Bikonta species. Sci. Rep. **6**, 31645 (2016)

K.W. Bender, W.A. Snedden, Calmodulin-related proteins step out from the shadow of their namesake. Plant Physiol. **163**, 486–495 (2013)

K.W. Bender, R.E. Zielinski, S.C. Huber, Revisiting paradigms of Ca(2+) signaling protein kinase regulation in plants. Biochem. J. **475**, 207–223 (2018)

N. Bouche, A. Yellin, W.A. Snedden, H. Fromm, Plant-specific calmodulin-binding proteins. Annu. Rev. Plant Biol. **56**, 435–466 (2005)

M. Boudsocq, J. Sheen, CDPKs in immune and stress signaling. Trends Plant Sci. **18**, 30–40 (2013)

L. Chae, G.K. Pandey, S. Luan, Y.H. Cheong, K.N. Kim, Protein Kinases and Phosphatases for Stress Signal Transduction in Plants, in *Abiotic Stress Adaptation in Plants*, ed. by A. Pareek, S. K. Sopory, H. J. Bohnert, (Springer, Dordrecht, 2010), pp. 123–163

M. Charpentier, Calcium signals in the plant nucleus: Origin and function. J. Exp. Bot **62**, 4165–4173 (2018)

R. Das, G.K. Pandey, Expressional analysis and role of calcium regulated kinases in abiotic stress Signaling. Curr. Genomics **11**, 2–13 (2010)

T.A. DeFalco, K.W. Bender, W.A. Snedden, Breaking the code: Ca^{2+} sensors in plant signalling. Biochem. J. **425**, 27–40 (2010)

A.N. Dodd, J. Kudla, D. Sanders, The language of calcium signaling. Annu. Rev. Plant Biol. **61**, 593–620 (2010)

W. Du, H. Lin, S. Chen, Y. Wu, J. Zhang, A.T. Fuglsang, M.G. Palmgren, W. Wu, Y. Guo, Phosphorylation of SOS3-like calcium-binding proteins by their interacting SOS2-like protein kinases is a common regulatory mechanism in Arabidopsis. Plant Physiol. **156**, 2235–2243 (2011)

K.H. Edel, E. Marchadier, C. Brownlee, J. Kudla, A.M. Hetherington, The evolution of calcium-based signalling in plants. Curr. Biol. **27**, R667–r679 (2017)

J.L. Gifford, M.P. Walsh, H.J. Vogel, Structures and metal-ion-binding properties of the Ca^{2+}-binding helix-loop-helix EF-hand motifs. Biochem. J. **405**, 199–221 (2007)

D. Gong, Y. Guo, A.T. Jagendorf, J.K. Zhu, Biochemical characterization of the Arabidopsis protein kinase SOS2 that functions in salt tolerance. Plant Physiol. **130**, 256–264 (2002)

L.P. Hamel, J. Sheen, A. Seguin, Ancient signals: Comparative genomics of green plant CDPKs. Trends Plant Sci. **19**, 79–89 (2014)

J.F. Harper, M.R. Sussman, G.E. Schaller, C. Putnam-Evans, H. Charbonneau, A.C. Harmon, A calcium-dependent protein kinase with a regulatory domain similar to calmodulin. Science **252**, 951–954 (1991)

K. Hashimoto, J. Kudla, Calcium decoding mechanisms in plants. Biochimie **93**, 2054–2059 (2011)

K. Hashimoto, C. Eckert, U. Anschutz, M. Scholz, K. Held, R. Waadt, A. Reyer, M. Hippler, D. Becker, J. Kudla, Phosphorylation of calcineurin B-like (CBL) calcium sensor proteins by their CBL-interacting protein kinases (CIPKs) is required for full activity of CBL-CIPK complexes toward their target proteins. J. Biol. Chem. **287**, 7956–7968 (2012)

E.M. Hrabak, C.W. Chan, M. Gribskov, J.F. Harper, J.H. Choi, N. Halford, J. Kudla, S. Luan, H.G. Nimmo, M.R. Sussman, M. Thomas, K. Walker-Simmons, J.K. Zhu, A.C. Harmon, The Arabidopsis CDPK-SnRK superfamily of protein kinases. Plant Physiol. **132**, 666–680 (2003)

S. Huang, R. Waadt, M. Nuhkat, H. Kollist, R. Hedrich, M.R.G. Roelfsema, Calcium signals in guard cells enhance the efficiency by which abscisic acid triggers stomatal closure. New Phytol. **224**, 177–187 (2019)

M. Ikura, J.B. Ames, Genetic polymorphism and protein conformational plasticity in the calmodulin superfamily: Two ways to promote multifunctionality. Proc. Natl. Acad. Sci. U. S. A. **103**, 1159–1164 (2006)

T. Ito, M. Nakata, J. Fukazawa, S. Ishida, Y. Takahashi, Alteration of substrate specificity: The variable N-terminal domain of tobacco Ca2+−dependent protein kinase is important for substrate recognition. Plant Cell **22**, 1592–1604 (2010)

T. Ito, S. Ishida, S. Oe, J. Fukazawa, Y. Takahashi, Autophosphorylation affects substrate-binding affinity of tobacco Ca2+−dependent protein Kinase1. Plant Physiol. **174**, 2457–2468 (2017)

T.J. Kleist, A.L. Spencley, S. Luan, Comparative phylogenomics of the CBL-CIPK calcium-decoding network in the moss Physcomitrella, Arabidopsis, and other green lineages. Front. Plant Sci. **5**, 187 (2014)

R.H. Kretsinger, C.E. Nockolds, Carp muscle calcium-binding protein. II. Structure determination and general description. J. Biol. Chem. **248**, 3313–3326 (1973)

V. La Verde, P. Dominici, A. Astegno, Towards understanding plant calcium signaling through Calmodulin-like proteins: A biochemical and structural perspective. Int. J. Mol. Sci. **19**, 1331 (2018)

S. Luan, The CBL-CIPK network in plant calcium signaling. Trends Plant Sci. **14**, 37–42 (2009)

E. McCormack, J. Braam, Calmodulins and related potential calcium sensors of Arabidopsis. New Phytol. **159**, 585–598 (2003)

C.A. McPhalen, N.C. Strynadka, M.N. James, Calcium-binding sites in proteins: A structural perspective. Adv. Protein Chem. **42**, 77–144 (1991)

M. Ohta, Y. Guo, U. Halfter, J.K. Zhu, A novel domain in the protein kinase SOS2 mediates interaction with the protein phosphatase 2C ABI2. Proc. Natl. Acad. Sci. U. S. A. **100**, 11771–11776 (2003)

G.K. Pandey, Emergence of a novel calcium signaling pathway in plants: CBL-CIPK signaling network. Physiol. Mol. Biol. Plants **14**, 51–68 (2008)

G.K. Pandey, P. Kanwar, A. Pandey, *Global Comparative Analysis of CBL-CIPK Gene Families in Plants* (Springer, New York, 2014)

G.K. Pandey, P. Kanwar, A. Singh, L. Steinhorst, A. Pandey, A.K. Yadav, I. Tokas, S.K. Sanyal, B.G. Kim, S.C. Lee, Y.H. Cheong, J. Kudla, S. Luan, Calcineurin B-like protein-interacting protein kinase CIPK21 regulates osmotic and salt stress responses in Arabidopsis. Plant Physiol. **169**, 780–792 (2015)

T. Romeis, M. Herde, From local to global: CDPKs in systemic defense signaling upon microbial and herbivore attack. Curr. Opin. Plant Biol. **20**, 1–10 (2014)

T. Romeis, A.A. Ludwig, R. Martin, J.D. Jones, Calcium-dependent protein kinases play an essential role in a plant defence response. EMBO J. **20**, 5556–5567 (2001)

M.J. Sanchez-Barrena, M. Martinez-Ripoll, A. Albert, Structural biology of a major signaling network that regulates plant abiotic stress: The CBL-CIPK mediated pathway. Int. J. Mol. Sci. **14**, 5734–5749 (2013)

D. Sanders, J. Pelloux, C. Brownlee, J.F. Harper, Calcium at the crossroads of signaling. Plant Cell **14 Suppl**, S401–S417 (2002)

S.K. Sanyal, A. Pandey, G.K. Pandey, The CBL-CIPK signaling module in plants: A mechanistic perspective. Physiol. Plant. **155**, 89–108 (2015)

S.K. Sanyal, S. Rao, L.K. Mishra, M. Sharma, G.K. Pandey, Plant Stress Responses Mediated by CBL-CIPK Phosphorylation Network, in *The Enzymes*, ed. by C. Lin, S. Luan, (Elsevier, Amsterdam, 2016), pp. 31–64

S.K. Sanyal, S. Mahiwal, G.K. Pandey, Calcium Signaling: A Communication Network that Regulates Cellular Processes, in *Sensory Biology of Plants*, ed. by S. Sopory, (Springer, Singapore, 2019), pp. 279–309

S.K. Sanyal, S. Mahiwal, D.M. Nambiar, G.K. Pandey, CBL-CIPK module-mediated phospho-regulation: Facts and hypothesis. Biochem. J. **477**, 853–871 (2020)

A. Simeunovic, A. Mair, B. Wurzinger, M. Teige, Know where your clients are: Subcellular localization and targets of calcium-dependent protein kinases. J. Exp. Bot. **67**, 3855–3872 (2016)

A. Singh, A.K. Yadav, K. Kaur, S.K. Sanyal, S.K. Jha, J.L. Fernandes, P. Sharma, I. Tokas, A. Pandey, S. Luan, G.K. Pandey, Protein phosphatase 2C, AP2C1 interacts with and negatively regulates the function of CIPK9 under potassium deficient conditions in Arabidopsis. J. Exp. Bot **69**, 4003–4015 (2018)

W.A. Snedden, H. Fromm, Calmodulin as a versatile calcium signal transducer in plants. New Phytol. **151**, 35–66 (2001)

N.C. Strynadka, M.N. James, Crystal structures of the helix-loop-helix calcium-binding proteins. Annu. Rev. Biochem. **58**, 951–998 (1989)

Y. Takahashi, T. Ito, Structure and function of CDPK: A sensor responder of calcium, in *Coding and Decoding of Calcium Signals in Plants*, ed. by S. Luan, (Springer, Berlin, 2011), pp. 129–146

A.K. Yadav, S.K. Jha, S.K. Sanyal, S. Luan, G.K. Pandey, Arabidopsis calcineurin B-like proteins differentially regulate phosphorylation activity of CBL-interacting protein kinase 9. Biochem. J. **475**, 2621–2636 (2018)

S. Ying, A.T. Hill, M. Pyc, E.M. Anderson, W.A. Snedden, R.T. Mullen, Y.M. She, W.C. Plaxton, Regulatory phosphorylation of bacterial-type PEP carboxylase by the Ca(2+)-dependent protein kinase RcCDPK1 in developing Castor oil seeds. Plant Physiol. **174**, 1012–1027 (2017)

H. Zeng, L. Xu, A. Singh, H. Wang, L. Du, B.W. Poovaiah, Involvement of calmodulin and calmodulin-like proteins in plant responses to abiotic stresses. Front. Plant Sci. **6**, 600 (2015)

Chapter 5
Plant Ion Channels without Molecular Identity and Two-Pore Channel 1

Contents

Introduction

We have already mentioned in second chapter that the patch-clamp technique had fueled the research of plant ion channels. The presence of voltage regulated (and voltage-independent) ion channels have been confirmed at the plasma membrane (PM) and other endomembranes in several plant species (Swarbreck et al. 2013). The major bottleneck in their (these voltage-dependent and -independent) research is the failure to identify the genes that code for hyperpolarization-activated Ca^{2+} channel (HACC), depolarization-activated Ca^{2+} channel (DACC) or voltage insensitive Ca^{2+} channel (VICC) channels in plants. Nevertheless, there is a significant amount of data on the Ca^{2+} uptake by these channels that indicate the importance of these channels in plant Ca^{2+} uptake. Demidchik and Shabala, have indicated that the Ca^{2+} conductances mediated by these channels show difference in their- (1) Activation kinetics (slow activating, rapid activating or 'spiky' burst-like activating), (2) Pharmacological sensitivity (to organic blockers and Lanthanides (Gd^{3+} and La^{3+})), (3) selectivities (besides Ca^{2+} permeable to (good permeability to K^+, Na^+, NH_4^+, Cs^+, Rb^+, Ba^{2+} and Sr^{2+}) (weak permeability to Zn^{2+}, Mn^{2+} and Mg^{2+})), (4) single-channel conductance, and (5) natural (or synthetic) modulators sensitivity (e.g., activation by polyamines, ROS, neurotransmitters, elicitors) (Demidchik and Shabala 2018; Demidchik et al. 2018). These channels were majorly identified

© The Editor(s) (if applicable) and The Author(s), under exclusive license to Springer Nature Switzerland AG 2021
G. K. Pandey, S. K. Sanyal, *Functional Dissection of Calcium Homeostasis and Transport Machinery in Plants*, SpringerBriefs in Plant Science, https://doi.org/10.1007/978-3-030-58502-0_5

through electrophysiological studies and have contributed to our understanding of Ca^{2+} transport (uptake) into the plant (or plant cell) (Demidchik and Shabala 2018; Demidchik et al. 2018).

Depolarization Activated Ca^{2+} Channels (DACCs)

The DACCs have always posed a challenge to the plant scientists because of their instability and infrequent appearance in the plant cell (Miedema et al. 2008). Their activity is quickly lost in the patch-clamp experiment (Swarbreck et al. 2013). Yet they have been reported in algae, *Daucus carota* L., *Arabidopsis thaliana* and *Triticum aestivum* (Demidchik and Shabala 2018). The DACCs have slow activation kinetics and microtubule disruption or addition of brassinosteroids increases their activity (Thion et al. 1996; Straltsova et al. 2015). Demidchik and Shabala have indicated the importance of DACCs in plants- (1) as they operate in membrane potential of around −100 to +30 mV (depolarized PM), they are important player involved in cytosolic Ca^{2+} entry during stress (i.e., they help in the transient Ca^{2+} rise in the cytosol), (2) these channels can self-inactivate, blocking the further Ca^{2+} increase in cell and letting the cell recover for next signaling event (Demidchik and Shabala 2018). The DACCs are operational only at the PM in the plant cell and have been reported in roots, xylem and guard cells.

Hyperpolarization Activated Ca^{2+} Channel (HACC)

The HACCs operate in guard cell closure, polar growth, roots and are modulated by light, ABA, ROS, nanoparticles and other elicitors. HACCs allow large Ca^{2+} influx at these locations (especially elongation zone of the root, tip of root hairs and pollen tubes) so that Ca^{2+} mediated process can be functional. They have been identified in several plant species (Demidchik and Shabala 2018). The HACCs can show both slow and fast kinetics (Demidchik and Shabala 2018). Hyperpolarized voltages (−150 to −200 mV) at the PM and activators (ABA, jasmonic acid, ROS, pathogen elicitors and purines) modulate HACC for Ca^{2+} entry into the cytosol (Demidchik and Shabala 2018). Also, transient rise in cytosolic Ca^{2+} positively modulates the HACCs to release Ca^{2+} in the cytosol (this is probably the reason for HACCs mediating large Ca^{2+} flux into the cytosol). ABA and ROS play a critical role in HACC mediated Ca^{2+} influx into the stomata to regulate stomatal movements (opening and closing). Pharmacological studies performed on HACCs indicate that they are blocked by lanthanides (GD^{3+} has been used in DACC study as well) (Miedema et al. 2008). They also show some sensitivity to Ca^{2+} channel antagonists (like dihydropyridines, phenylalkylamines, benzothiazepine and bepridil). The HACCs are insensitive to K^+ and anion channel blockers. The HACC can transport divalent (Ca^{2+}, Ba^{2+} and Sr^{2+}) as well as monovalent cations (K^+, Na^+, Cs^+, Li^+, TEA^+) cations

(Demidchik and Shabala 2018). The HACCs are observed in the PM as well as vacuole in the plant cell.

Voltage Insensitive Ca^{2+} Channel (VICC)

The VICCs operate at the PM of the plant cell and are observed in the roots xylem and guard cells. They have faster kinetics (rapid activation) compared to DACC and HACC. VICCs have been reported in *Arabidopsis* and wheat (Straltsova et al. 2015). VICCs majorly maintain the Ca^{2+} levels at resting potentials and are hence thought to supply Ca^{2+} for plants nutritional and structural needs. VICCs can be activated by glutamate, purines and ROS. The ROS regulation of VICC and HACC are different and in the resting membrane potential of PM (about -130 to -100 mV), ROS activated VICCs are major Ca^{2+} influx elements in the roots (compared to ROS activated HACCs) (Demidchik and Shabala 2018). Again lanthanides can block their activity, and they are also sensitive to quinine and diethylpyrocarbonate (Demidchik et al. 2002). Similar to HACCs, they are insensitive to K$^+$ and anion channel blockers and also to organic Ca^{2+} channel blockers. The VICCs allow an influx of Na$^+$, Mg^{2+}, Zn^{2+} and Mn^{2+} into the cell (Demidchik and Shabala 2018). The VICCs are sole moderators of Ca^{2+} uptake in tissues where HACCs are absent.

Many researchers have speculated about the identities of the HACC, DACCs and VICCs. The hypothesis states that HACCs may be annexins, VICCs may be glutamate receptors (GLRs) or cyclic nucleotide-gated channels (CNGCs) and DACC may be two-pore channel 1 (TPC1) (White and Broadley 2003; Swarbreck et al. 2013). In this writeup, we have discussed DACC and TPC1 separately. Further, the TPC1 is currently considered as slow vacuolar (SV) channel and is discussed in the next section.

Two-Pore Channel1 (TPC1)

There is an important question regarding TPCs ability to mediate Ca^{2+} mediated Ca^{2+} release (CICR) (we have discussed CICR in Chap. 2). In this section, we will discuss the chief ambiguity regarding the TPC i.e., is it a Ca^{2+} transport channel involved in CICR or Ca^{2+} regulated cation (K$^+$, Na$^+$) channel, by looking at the current data (electrophysiology, structural, genetic combined with Ca^{2+} sensors and functional aspect). Rainer Hedrich and colleague have very elegantly articulated the beginning of TPC research (Hedrich and Marten 2011). The search for ion channels through electrophysiology had progressed gradually from whole-cell level to cell organelles. Vacuole being the largest organelle in the plant cell had received its due share of attention for the investigation of ion channels. Due to the concerted (and excellent) work by different research groups, different channels were identified, and among them, a slow activating (SV) channel showed large-conductance in presence of cytosolic Ca^{2+} ions (experiments done on sugarbeet) (Hedrich and Neher 1987). Many other researchers confirmed that SV channels were present in the vacuoles of other plant species as well. In the meantime, animal TPC1 were reported and were identified in several plants species (Hedrich and Marten 2011). It was proven experimentally that GFP tagged Arabidopsis TPC1 could localize to the vacuole and patch-clamp studies on the vacuoles isolated from *tpc1* knockout and TPC1 overexpression lines confirmed that TPC1 was indeed the earlier reported SV channel (Peiter et al. 2005; Hedrich and Marten 2011).

The Structure of TPC1 Channel

To understand the function of the TPC1 channel, we must understand how the channel functions at the structural level. The last decade has seen a major development in the crystal structure of TPC1, and Arabidopsis TPC1 structure has been reported by two different groups (Guo et al. 2016; Kintzer and Stroud 2016). Besides, the structure of mouse and human TPCs have also been determined (She et al. 2018, 2019). Our discussion here is mainly on Arabidopsis TPC1 (mentioned just TPC1).

The most important statement about plant TPC1 is that they are "voltage-gated in a Ca^{2+} dependent manner" and "non-selective" ion channels (Patel et al. 2016). We now look into the structural data to understand the relevance of these two terms. A single TPC polypeptide contains two tandem shakers like domain (this nomenclature is from other voltage-gated ion channels with a similar architecture) named D1 and D2. Each domain has 6 transmembrane helices (named IS1 to IS6 and IIS1 to IIS6) and a linker connecting D1 and D2 has two EF-hands (EF1 and EF2). Each domain of 6 helices can be further classified into voltage-sensing domain [(IS1 to IS4 (VSD1) and IIS1 to IIS4 (VSD2)); S1 to S12 in some literature] and pore helices (P1 between IS5 to IS6 and P2 between IIS5 to IIS6). Then there are the N-terminal domain (NTD) and C-terminal domain (CTD). The TPC1 dimer assembles at the vacuolar membrane to form a functional channel with the NTD, EF1, EF2 and CTD being on the cytoplasmic side, the helices at the membrane and some connecting regions (of the transmembrane helices) at the vacuolar lumen side. NTD ensures vacuolar targeting of TPC, and removal of either NTD or CTD makes the channel dysfunctional (Guo et al. 2016; Kintzer and Stroud 2016; Hedrich et al. 2018; Kintzer and Stroud 2018; She et al. 2018, 2019).

Now we first look into the question of Ca^{2+} binding. The TPC has two distinct Ca^{2+} binding sites, one at the cytoplasmic with two EF-hands and the other in the vacuolar lumen (Kintzer and Stroud 2018). The function of Ca^{2+} sensing is different, the cytoplasmic Ca^{2+} sensing is done to open the channel for ion release into the cytosol, the vacuolar side Ca^{2+} sensing is done to inhibit the channel, so to stop the vacuolar Ca^{2+} from leaking into the cytoplasm. The EF-hands are obvious candidates for Ca^{2+} sensing in the cytosolic side, but there are contrasting reports on the Ca^{2+} binding property of the two EF-hands (based on their sequence, mutational studies and crystal structure) (Schulze et al. 2011; Guo et al. 2016; Kintzer and

Stroud 2016). A simplified view, combining different observations, is that Ca^{2+} binding to EF-hands will make some conformational changes in TPC1 that will help in opening the gates for ion conductance (Patel et al. 2016). At the vacuolar side, there are two different sites for Ca^{2+} binding. Again the two different reports do not exactly match on the amino acids involved, but it is clear that site 1 with Asp454 (D454) is the most important site for sensing vacuolar Ca^{2+} (Beyhl et al. 2009; Dadacz-Narloch et al. 2011; Guo et al. 2016; Kintzer and Stroud 2016). Ca^{2+} binding to site 1 clamps the S4 to IIS1-IIS2 loop thereby inhibiting TPC1 (Patel et al. 2016). Mutation in this site to Asn (N) in the fatty acid oxygenation upregulated 2 (*fou2*) mutant makes the D454NTPC1 insensitive to vacuolar Ca^{2+} and increases the rate of channel opening (Bonaventure et al. 2007a, b; Guo et al. 2016; Kintzer and Stroud 2016).

Now we look into "voltage sensing" of TPC1. The two VSDs (VSD1 and VSD2) each have its IS4 (IIS4 in VSD2), which are important for membrane voltage sensing (Guo et al. 2016; Kintzer and Stroud 2016; She et al. 2018, 2019). In the classical "voltage sensor," the helix will have three to eight Arg or Lys residues at every third position interspersed with predominantly hydrophobic residues (Hedrich et al. 2018). The IS4 of TPC1 only has 2 Arg and forms a regular helix (for voltage sensing needs to form a 3_{10}-helix motif seen in other voltage-gated channels) (Guo et al. 2016; Kintzer and Stroud 2016). Also, important residues forming the charge transfer center (another important feature of voltage-gated channels) in IS2 are replaced. These facts indicate that VSD1 is not the voltage-sensing domain in TPC1 (Guo et al. 2016; Kintzer and Stroud 2016). All these features are present in VSD2 (IIS4 has 4 conserved Arg and is a 3_{10}-helix) and thus is the voltage sensing center of TPC1 (Guo et al. 2016; Jaślan et al. 2016; Kintzer and Stroud 2016; She et al. 2018). The human TPC2 does not have some key features in its IS4 and IIS4 making it voltage insensitive (She et al. 2019). Human TPC1 involves both VSD1 and VSD2 in voltage sensing and mouse TPC1 involves only VSD2 (She et al. 2018). Thus for activation of TPC1, both cytosolic Ca^{2+} and membrane voltage are required (in animals, it is phosphatidylinositol 3,5-bisphosphate ($PI(3,5)P_2$) and voltage) (Hedrich et al. 2018).

Finally, we come to the pore region responsible for the ion selectivity of an ion channel. This region determines the ions that can pass through a channel and are therefore designated as 'selectivity filters' (Guo et al. 2017). As stated earlier, TPC1 is non-selective with permeability being $Ca^{2+} > Na^+$, Li^+, K^+ (almost similar for three) $> Rb^+$. Cs^+ (in comparison human TPC1 is Na^+ selective and human TPC2 transports Na^+, Ca^{2+} and possibly H^+) (experiments on TPC1 conductivity performed in non-homologous system) (Guo et al. 2017). The P1 of TPC1 has residues Thr264Ser265 and P2 has residues Met629Gly630Asn631 as filter residues, which makes it a non-selective channel. When the Ser265, Met629 and Gly630 were changed to mimic the human TPC2 filter, TPC1 became selective for Na^+ only (Guo et al. 2017). The *fou2* mutant, had insensitivity towards luminal Ca^{2+} and showed a hypersensitive phenotype. In *ouf8* mutant, this is reversed (the *ouf8* TPC1 becomes sensitive to luminal Ca^{2+} like wild type TPC) due to a mutation in Met629, which was changed to Ile thereby blocking the TPC1 from releasing K^+ into the cytosol

(Lenglet et al. 2017). The structure and the sequences are represented in Figs. 5.1 and 5.2.

Ca²⁺ Conductance, CICR and SV Channel

The plant SV channels can transport K^+ (K^+ homeostasis) and Na^+ (during Na^+ cycling in salt stress) (Hedrich et al. 2018; Shabala et al. 2020). It was also shown that it could transport Ca^{2+}, however, the Ca^{2+} gradient used for the study was above

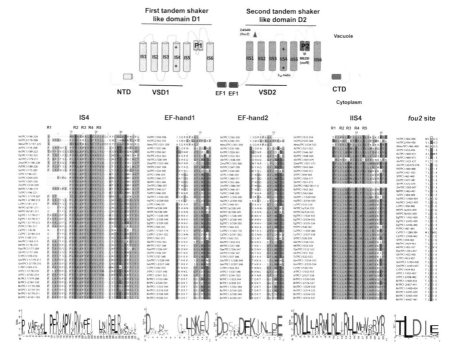

Fig. 5.1 Structure of plant TPC1 (based on Arabidopsis TPC1). The tandem domain structure of TPC1 is presented. The domains D1 and D2 are coloured differently. Each domain has 6 transmembrane helices, and S4 helix serves as the voltage-sensor (in TPC1 only IIS4 is functional and forms a 3_{10}-helix). The cytoplasmic side has the NTD (N-terminal domain), two tandem EF-hands and the CTD (C-terminal domain). The location of the *fou2* and *ouf8* mutations are marked in the structure. The structure is according to (Guo et al. 2016) and location of the mutations are according to (Lenglet et al. 2017). The multiple sequence alignment of IS4, IIS4, EF-hand1 and EF-hand2 and fatty acid oxygenation upregulated 2 (*fou2*) are represented. Note IS4 has smaller number of critical Arg (R) and hence are considered not to play any role in voltage-sensing. The IIS4 has 4 critical Arg and so are the main site for voltage-sensing. Both the EF-hands differ from canonical sequence, but EF-hand2 has more residues, which can co-ordinate Ca^{2+} (detail on EF-hands refer Chap. 4). The *fou2* site is very conserved in plant species, indicating a similar mechanism for sensing vacuolar Ca^{2+}. The sequences taken from UniProt database and MSA performed by MEGA10 and viewed using Jalview. The consensus was viewed using WebLogo

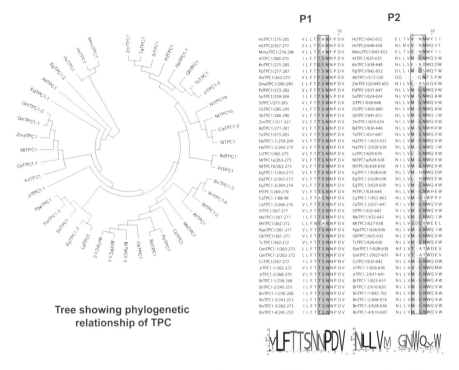

Fig. 5.2 Phylogenetic relationship of TPCs based on their pore sequence. A maximum-likelihood tree is presented rooted to the human TPC (1 and 2) and mouse TPC1 (the three marked in red). The plants TPC form a different clade depending on their pore sequence, which is different than their animal counterpart. The analyzed sequences majorly indicate that they are non-selective compared to animals, which are majorly Na⁺ selective. Multiple sequence alignment of the pore region (P1 and P2) is provided (according to (Guo et al. 2017)). The sequences taken from UniProt database and MSA performed by MEGA10 and viewed using Jalview. The consensus was viewed using WebLogo. Tree generated using MEGA10

physiological limits (50 mM CaCl$_2$ in the vacuolar lumen and 5 mM CaCl$_2$ with 100 mM KCl in the cytoplasmic side) (Ward and Schroeder 1994). It was also proven by a separate group that SV channels contributed to CICR (Bewell et al. 1999) but a contradictory report suggests TPC1 does not generate transient Ca²⁺ rise during abiotic or biotic stress (Ranf et al. 2008). However, it was also shown that in the *attpc1–2* mutant, the stomata do not respond to extracellular Ca²⁺, indicating its involvement in CICR (Peiter et al. 2005). Also, separate group showed that *tpc1–2* mutants could still elicit a Ca²⁺ wave (cytosolic Ca²⁺ rise) but at a slower speed compared to when TPC1 was present (Choi et al. 2014). Another group has reported that TPC1 is involved in wounding related Ca²⁺ wave generation (Vincent et al. 2017). In the heterologous system, TPC1 could transport Ca²⁺ and further, it has been recently shown that the TPC1 and K⁺ transport channel l (TPK1) in the vacuole are responsible for electrical excitability of the vacuolar membrane (Jaślan et al. 2019). Also, TPC1 could modulate anion channels during stomatal closure in response to elevated cytosolic Ca^{2+} (Islam et al. 2010). Summarizing all the current information, we can conclude that the SV channels found by electrophysiological data are the TPC1 channel. They are Ca^{2+} regulated non-selective channels taking part in CICR in plants in response to stimuli. Their CICR mediated by TPC could be in turn modulated by ROS generated Ca^{2+} waves (we have talked about ROS regulating Ca^{2+} signaling in Chap. 3) (Hedrich et al. 2018).

TPC1 in Other Plants and their Functional Conservation

TPCs are considered eukaryotic gene as its orthologues are absent in prokaryotes (Hedrich et al. 2018). TPC is found in charophyte algae again proving that some of the Ca^{2+} signaling elements were an integral part of land colonization of marine algae. Voltage-dependent gating mechanism and Ca^{2+} sensing (cytosolic as well as vacuolar) appears to be fairly similar for plant TPCs, only the pore region (ion selectivity filter) may be different, in Fig. 5.2 (Hedrich et al. 2018). Hedrich group has indicated that *Physcomitrella patens* TPC1 can generate 'archetypical' SV currents in *tpc1–2* mutants further indicating that the TPC1 in plants may be functionally conserved (Hedrich et al. 2018). Rice and wheat TPC1 also show similar property (Dadacz-Narloch et al. 2013).

Conclusion

The HACC, DACC and VICC contributed significantly to our knowledge of Ca^{2+} uptake by plants. But they have yet eluded identification, and that has hindered further decisive experimentations. The TPC1 channels have been a big breakthrough in this regard. Its crystal structure has aided our understanding of the mechanism that is employed to modulate Ca^{2+} homeostasis inside the plant cell.

References

M.A. Bewell, F.J. Maathuis, G.J. Allen, D. Sanders, Calcium-induced calcium release mediated by a voltage-activated cation channel in vacuolar vesicles from red beet. FEBS Lett. **458**, 41–44 (1999)

D. Beyhl, S. Hörtensteiner, E. Martinoia, E.E. Farmer, J. Fromm, I. Marten, R. Hedrich, The fou2 mutation in the major vacuolar cation channel TPC1 confers tolerance to inhibitory luminal calcium. Plant J. **58**, 715–723 (2009)

G. Bonaventure, A. Gfeller, V.M. Rodríguez, F. Armand, E.E. Farmer, The fou2 gain-of-function allele and the wild-type allele of two pore channel 1 contribute to different extents or by different mechanisms to defense gene expression in Arabidopsis. Plant Cell Physiol. **48**, 1775–1789 (2007a)

G. Bonaventure, A. Gfeller, W.M. Proebsting, S. Hörtensteiner, A. Chételat, E. Martinoia, E.E. Farmer, A gain-of-function allele of TPC1 activates oxylipin biogenesis after leaf wounding in Arabidopsis. Plant J. **49**, 889–898 (2007b)

W.G. Choi, M. Toyota, S.H. Kim, R. Hilleary, S. Gilroy, Salt stress-induced Ca2+ waves are associated with rapid, long-distance root-to-shoot signaling in plants. Proc. Natl. Acad. Sci. U. S. A. **111**, 6497–6502 (2014)

B. Dadacz-Narloch, D. Beyhl, C. Larisch, E.J. López-Sanjurjo, R. Reski, K. Kuchitsu, T.D. Müller, D. Becker, G. Schönknecht, R. Hedrich, A novel calcium binding site in the slow Vacuolar cation channel TPC1 senses luminal calcium levels. Plant Cell **23**, 2696–2707 (2011)

B. Dadacz-Narloch, S. Kimura, T. Kurusu, E.E. Farmer, D. Becker, K. Kuchitsu, R. Hedrich, On the cellular site of two-pore channel TPC1 action in the Poaceae. New Phytol. **200**, 663–674 (2013)

V. Demidchik, S. Shabala, Mechanisms of cytosolic calcium elevation in plants: The role of ion channels, calcium extrusion systems and NADPH oxidase-mediated 'ROS-Ca(2+) Hub'. Funct. Plant Biol. **45**, 9–27 (2018)

V. Demidchik, H.C. Bowen, F.J. Maathuis, S.N. Shabala, M.A. Tester, P.J. White, J.M. Davies, Arabidopsis thaliana root non-selective cation channels mediate calcium uptake and are involved in growth. Plant J. **32**, 799–808 (2002)

V. Demidchik, S. Shabala, S. Isayenkov, T.A. Cuin, I. Pottosin, Calcium transport across plant membranes: Mechanisms and functions. New Phytol. **220**, 49–69 (2018)

J. Guo, W. Zeng, Q. Chen, C. Lee, L. Chen, Y. Yang, C. Cang, D. Ren, Y. Jiang, Structure of the voltage-gated two-pore channel TPC1 from Arabidopsis thaliana. Nature **531**, 196–201 (2016)

J. Guo, W. Zeng, Y. Jiang, Tuning the ion selectivity of two-pore channels. Proc. Natl. Acad. Sci. U. S. A. **114**, 1009–1014 (2017)

R. Hedrich, I. Marten, TPC1-SV channels gain shape. Mol. Plant **4**, 428–441 (2011)

R. Hedrich, E. Neher, Cytoplasmic calcium regulates voltage-dependent ion channels in plant vacuoles. Nature **329**, 833–836 (1987)

R. Hedrich, T.D. Mueller, D. Becker, I. Marten, Structure and function of TPC1 Vacuole SV channel gains shape. Mol. Plant **11**, 764–775 (2018)

M.M. Islam, S. Munemasa, M.A. Hossain, Y. Nakamura, I.C. Mori, Y. Murata, Roles of AtTPC1, vacuolar two pore channel 1, in Arabidopsis stomatal closure. Plant Cell Physiol. **51**, 302–311 (2010)

D. Jaślan, T.D. Mueller, D. Becker, J. Schultz, T.A. Cuin, I. Marten, I. Dreyer, G. Schönknecht, R. Hedrich, Gating of the two-pore cation channel AtTPC1 in the plant vacuole is based on a single voltage-sensing domain. Plant Biol. (Stuttg.) **18**, 750–760 (2016)

D. Jaślan, I. Dreyer, J. Lu, R. O'Malley, J. Dindas, I. Marten, R. Hedrich, Voltage-dependent gating of SV channel TPC1 confers vacuole excitability. Nat. Commun. **10**, 1–9 (2019)

A.F. Kintzer, R.M. Stroud, Structure, inhibition and regulation of two-pore channel TPC1 from Arabidopsis thaliana. Nature **531**, 258–262 (2016)

A.F. Kintzer, R.M. Stroud, On the structure and mechanism of two-pore channels. FEBS J. **285**, 233–243 (2018)

A. Lenglet, D. Jaślan, M. Toyota, M. Mueller, T. Müller, G. Schönknecht, I. Marten, S. Gilroy, R. Hedrich, E.E. Farmer, Control of basal jasmonate signalling and defence through modulation of intracellular cation flux capacity. New Phytol. **216**, 1161–1169 (2017)

H. Miedema, V. Demidchik, A.A. Véry, J.H. Bothwell, C. Brownlee, J.M. Davies, Two voltage-dependent calcium channels co-exist in the apical plasma membrane of Arabidopsis thaliana root hairs. New Phytol. **179**, 378–385 (2008)

S. Patel, C.J. Penny, T. Rahman, Two-pore channels enter the atomic era: Structure of plant TPC revealed. Trends Biochem. Sci. **41**, 475–477 (2016)

E. Peiter, F.J. Maathuis, L.N. Mills, H. Knight, J. Pelloux, A.M. Hetherington, D. Sanders, The vacuolar Ca2+−activated channel TPC1 regulates germination and stomatal movement. Nature **434**, 404–408 (2005)

S. Ranf, P. Wünnenberg, J. Lee, D. Becker, M. Dunkel, R. Hedrich, D. Scheel, P. Dietrich, Loss of the vacuolar cation channel, AtTPC1, does not impair Ca2+ signals induced by abiotic and biotic stresses. Plant J. **53**, 287–299 (2008)

C. Schulze, H. Sticht, P. Meyerhoff, P. Dietrich, Differential contribution of EF-hands to the Ca^{2+}-dependent activation in the plant two-pore channel TPC1. Plant J. **68**, 424–432 (2011)

S. Shabala, G. Chen, Z.-H. Chen, I. Pottosin, The energy cost of the tonoplast futile sodium leak. New Phytol. **225**, 1105–1110 (2020)

J. She, J. Guo, Q. Chen, W. Zeng, Y. Jiang, X. Bai, Structural insights into the voltage and phospholipid activation of mammalian TPC1 channel. Nature **556**, 130–134 (2018)

J. She, W. Zeng, J. Guo, Q. Chen, X. Bai, Y. Jiang, Structural mechanisms of phospholipid activation of the human TPC2 channel. eLife **8**, e45222 (2019)

D. Straltsova, P. Chykun, S. Subramaniam, A. Sosan, D. Kolbanov, A. Sokolik, V. Demidchik, Cation channels are involved in brassinosteroid signalling in higher plants. Steroids **97**, 98–106 (2015)

S.M. Swarbreck, R. Colaco, J.M. Davies, Plant calcium-permeable channels. Plant Physiol. **163**, 514–522 (2013)

L. Thion, C. Mazars, P. Thuleau, A. Graziana, M. Rossignol, M. Moreau, R. Ranjeva, Activation of plasma membrane voltage-dependent calcium-permeable channels by disruption of microtubules in carrot cells. FEBS Lett. **393**, 13–18 (1996)

T.R. Vincent, M. Avramova, J. Canham, P. Higgins, N. Bilkey, S.T. Mugford, M. Pitino, M. Toyota, S. Gilroy, A.J. Miller, S.A. Hogenhout, D. Sanders, Interplay of plasma membrane and Vacuolar ion channels, together with BAK1, elicits rapid cytosolic calcium elevations in Arabidopsis during Aphid Feeding. Plant Cell **29**, 1460–1479 (2017)

J.M. Ward, J.I. Schroeder, Calcium-activated K+ channels and calcium-induced calcium release by slow vacuolar ion channels in guard cell vacuoles implicated in the control of stomatal closure. Plant Cell **6**, 669–683 (1994)

P.J. White, M.R. Broadley, Calcium in plants. Ann. Bot. **92**, 487–511 (2003)

Chapter 6
Plant Ligand-Gated Ion Channels 1: GLR

Contents

Introduction

Besides the voltage-gated ion channels discussed in the previous chapter, plants are equipped with another group of channels that mainly use ligands to modulate their channel activities (opening or closing of the channel). In this chapter, we talk about the plant glutamate receptor-like (GLR) ion channels.

The Glutamate Receptor-like (GLR) Channels

The ionotropic glutamate receptors (iGluR) respond to glutamate and are well characterized in the animal system (Weiland et al. 2016; Wudick et al. 2018a). These were later identified in the plant system (*Arabidopsis thaliana*) and their homology to the animal iGluR (henceforth animal glutamate receptors will be called iGluR) led researchers to name them as glutamate receptor-like (GLR, henceforth GLR means plant glutamate receptor-like) (Lam et al. 1998; Lacombe et al. 2001; Turano et al. 2001). The iGluR and the GLR are both non-selective ion channels, which are considered to be derivative of the eukaryotic voltage- and ligand-gated channel (discussed in the previous chapter) (Price et al. 2012). The GLR is now accepted as plant Ca^{2+} transporters although there are some ambiguities regarding its ligand gating, ion selectivity and subcellular localization (Wudick et al. 2018a).

G. K. Pandey, S. K. Sanyal, *Functional Dissection of Calcium Homeostasis and Transport Machinery in Plants*, SpringerBriefs in Plant Science,
https://doi.org/10.1007/978-3-030-58502-0_6

The GLR (or iGluR) are present in all life form except in yeast, eubacteria, archaebacteria and fungi (Wudick et al. 2018a). The iGluR are divided into four different classes (N-methyl-D-aspartate (NMDA), α-amino-3-hydroxy-5-methyl-4-isoxazolepropionic acid (AMPA), kainate receptors and δ-receptors (De Bortoli et al. 2016; Wudick et al. 2018a). These clades are different in their biochemical properties (like activation kinetics, ion selectivity). The Arabidopsis GLRs form three distinct clades quite different from the iGluRs (Chiu et al. 2002). In the next section, we discuss our current understanding of the structure of GLRs and their functional significance.

The GLR Structure

Unlike the case of plant TPC1, whose crystal structure was the first to be reported (compared to animals), in case of GLRs, the scenario is quite opposite. The reports on iGluR structure dominate the literature, and mostly due to sequence homology, the function of GLR domains are interpreted from iGluR (Chiu et al. 2002; Sobolevsky 2015; Zhu and Gouaux 2017). There is only one report on the ligand-binding domain (LBD) of Arabidopsis GLR (there is another submitted to the repository, at the time of writing, and since not peer-reviewed has not been considered for this write-up) (Alfieri et al. 2020). So, our discussion will majorly focus on the iGluR and then we will interpret the GLR function from that data.

In the membrane, a functional GLR will be a tetramer composed of four different GLR subunit (either homodimerization or heteromerization) (Turano et al. 2002; Vincill et al. 2012, 2013; Price et al. 2013; Sobolevsky 2015; Weiland et al. 2016). The multimerization step probably happens in the endoplasmic reticulum (ER) (Weiland et al. 2016). The single subunit of GLR can be divided into the amino-terminal domain (ATD), the S1 domain, followed by three transmembrane (henceforth TM) helices, then S2 domain, again a TM helix and lastly the C-terminal domain (CTD) (Weiland et al. 2016; Wudick et al. 2018a). The S1 and S2 combine to form the LBD. The ATD and LBD both have "clamshell" like structure facilitating substrate binding (Wudick et al. 2018a). The TM helices are named differently in literature, ranging from M1 to M4 named in order from the ATD or M1, pore, M2 and M3 (i.e., the small TM segment between first and second TM helix is named M2 or pore). For our discussion, we have considered this as pore (P) and hence follow the later nomenclature. Finally, the ATD and LBD are faced at opposite ends of the CTD. The GLR channel functions in the following paradigm- ligand binding to LBD causes conformational changes in the TM domain and hence the channel opens. In the case of iGluRs, after this, the channel closes and stays insensitive to ligand binding (desensitization phase) (Wudick et al. 2018a). Figure 6.1 denotes the structure of a single subunit of GLR and the functional tetrameric structure that is formed by the association of four GLR subunits.

Just before the ATD, there is N-terminal signal sequence (a transmembrane anchor) in iGluRs for guiding the subunit to the endoplasmic reticulum (ER). This

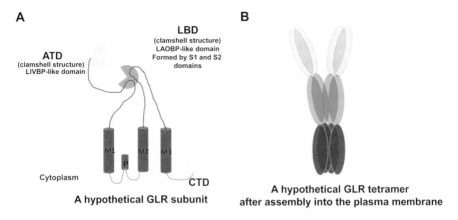

Fig. 6.1 Hypothetical structure of plant GLR. (a). The plant GLR has the ATD domain followed LBD domain (formed by S1 and S2 domains). The substrates can bind to the LBD domain to regulate the channel gating. Both ATD and LBD have a clamshell-like structure. These two domains are followed by the transmembrane helices M1, M2 and M3 and pore region between M1 and M2. (**b**). This is the hypothetical arrangement of the GLR tetramer (it can be a homomer or heteromer). This is the functional channel. The structure of single subunit is adopted and modified from (Wudick et al. 2018a), and the tetramer structure from (Sobolevsky 2015)

sequence is cleaved once the subunit reaches the ER (Wudick et al. 2018a). The Arabidopsis GLRs may have similar sequences which could guide them to ER by a similar mechanism as the iGluRs (Wudick et al. 2018a). This is followed by the ATD, which resemble a bacterial Leu/Ile/Val-binding protein (LIVBP) domains (Acher and Bertrand 2005). From studying the iGluRs, it is known that ATD is regulated by polyamines, Zn^{2+} and ifenprodil (information from NMDA receptors, for others it is unknown) (Wudick et al. 2018a). The ATDs may also help in the multimerization of iGluR in ER (information from NMDA, AMPA and kainite receptors) and iGluR gating (especially in the desensitization phase) (Wudick et al. 2018a). The N-terminus of plant GLRs (including the ATDs) are more similar to G-protein coupled receptors (Turano et al. 2001). As such they may have a different mode of activation (i.e., they may bind to amino acids like Gly) than the iGluRs (Wudick et al. 2018a). Further, desensitization (reports on GLR are ambiguous) and iGluR like conformational change have not been reported for plants. So, there are possibilities that this domain can function differently than their animal counterparts.

As mentioned earlier, the structure of LBD of plants (Arabidopsis GLR3.3, henceforth only GLR3.3) has been recently solved and here we model our discussion on its structure (Alfieri et al. 2020). There is an actual ligand-binding site in the GLRs (or iGluRs), and it is homologous to Lys/Arg/Orn-binding protein (LAOBP)-like domains from bacterial proteins. Like all LBD, the GLR3.3 is also bilobed. The lobes have two domains domain 1 (henceforth D1- with 6 α helices and 2 β strands) and domain 2 (henceforth D2- with 5 α helices and 5 β strands). The D1 domain has two loops loop1 and loop2 (we will discuss their relevance a little later). Between the D1 and D2, there is a deep cleft where the ligand binds. Here certain conserved amino acids help in the interaction with the ligands (some GLRs may have different amino acids here and hence different affinities for the same ligands). For example, GLR1 prefers D-Ser, GLR3.3 (Glu, Ala, Asn, Ser) and GLR3.4 prefers Glu. Also, the ligands themselves have additional interaction inside the cleft thus influencing the ligand-gated Ca^{2+} release by GLRs. When compared with iGluRs, the loop1 of GLR3.3 is expanded and may influence allosteric modulation of the tetrameric GLR subunits. The β1-αA and αH- β6 loops are expanded and loop2 is rearranged. The β1-αA and loop2 host the ligand interacting side chain so it indicates that ligand interaction should vary between GLRs and iGluRs due to this change (Alfieri et al. 2020). A comprehensive list of agonist and antagonist of GLR is presented in Table 6.1.

The next segment in the channel is the gate and the pore region that allow the ion flux. The gate keeps the pore locked and receptor binding causes large changes in the channels (from NTD domain twist to large rearrangement in the channel tetramers) and results in channel opening (Sobolevsky 2015; Twomey and Sobolevsky 2018). From the iGluR structure, the M1 to M3 TM helices form the core of the ion channel. In the multimeric form, the TM helices are arranged in symmetry with the core of the ion channel. P region will form the inner cavity of the pore in the multimer and the M2 lines the outer cavity. The M1 helix is placed exterior to pore and M2 and M3 of one subunit associates with (M1-Pore-M2) of another subunit (Traynelis et al. 2010). The M2 helix in iGluRs has a conserved "SYTANLAA" motif with second Ala considered responsible for gate opening (Wudick et al. 2018a). The pore region is composed of 20 amino acids and the iGluR are generally non-selective (as they do not possess typical residues that are responsible for K^+ selectivity of prokaryotic iGluRs) (Wudick et al. 2018a). The pore region of some of the iGluRs is subject to RNA editing that affects their ion selectivity (Traynelis et al. 2010). RNA editing is also seen in other sites (than the pore), which affects the general channel properties (Traynelis et al. 2010). Arabidopsis has significant differences in the SYNTANLAA motifs and hence there are significant differences in their gating properties. The clade 3 Arabidopsis GLRs (we discuss more on the clades in the next section) are closer to this motif and clade 1 and clade 3 show significant variation from this motif. GLR3.2 and GLR3.3 are active without ligands. The P region of iGluRs and GLRs are highly divergent and hence the similarities in ion selectivity cannot be directly inferenced by sequence comparison (Wudick et al.

Table 6.1 Representing the ligands tested for Arabidopsis GLRs

Arabidopsis GLR protein	Agonist	Antagonist	No effect
GLR1.4	Asn, D-Ser, Met, Trp, Tyr, Thr, Leu, Phe,	DNQX, CNQX, MK-801, Philanthotoxin	
GLR3.3	L-Glu, Gly, Asn, Ala, (L and D)Ser, Cys, Glutathione	DNQX, AP-5	D-Glu, GABA, NMDA, D-Ser (*)
GLR3.4	Gly, Asn, Ala, (L and D)Ser, Cys		L-Ala, L-Glu, D-Ala, Phe

The data is from (Weiland et al. 2016). For D-Ser mild activity was detected (Alfieri et al. 2020)

2018a) (in Fig. 6.2). Heterologus electrophysiological studies on many GLR1 (mostly Arabidopsis and one on *Physcomitrella patens*) show they are non-selective cation conductors with a preference for Ca^{2+} (Tapken and Hollmann 2008; Vincill et al. 2012; Tapken et al. 2013; Ortiz-Ramírez et al. 2017; Wudick et al. 2018b). Some *in planta* mutant studies and Ca^{2+} sensor studies also indicate their preference for Ca^{2+} conductance (Kim et al. 2001; Michard et al. 2011).

Some part of the S2 domain and the region preceding the M3 may have some unexplained role. The M3 helix is important for post-translational modification, gating and desensitization of iGluRs (Wudick et al. 2018a). The CTD in iGluR has some motifs for ER retention and some protein interaction motifs (Wudick et al. 2018a). Some contrasting reports on a similar mechanism have been reported which makes it hard to conclude if desensitization happens in plants and the GLR M3 helix play a similar role (Stephens et al. 2008; Vincill et al. 2012; Tapken et al. 2013). The GLRs have 14–3-3 interaction domain in their CTD and this made Wudick and colleagues hypothesize that the GLR CTD can be phosphorylated by Ser/Thr kinases to help in 14–3-3 interaction (Wudick et al. 2018a).

Evolution, Localization and Biological Function of GLRs

The initial classification of Arabidopsis GLRs has divided them into three clades based on their sequences and the nomenclature, following the name of the clades (e.g., clade 1 members were named GLR1 followed by another number) (Lacombe et al. 2001; Chiu et al. 2002). According to De-Bortoli and colleagues, the GLRs (iGluR bacterial iGLuR0) are thought to have evolved from the insertion of KsCA like K^+ channel with inverted membrane topology into a periplasmic binding protein (De Bortoli et al. 2016). The plant GLRs have significantly increased in number during evolution in the vascular plants, comprising very high number from lower to higher land plants (Arabidopsis has 20 GLRs compared to 2 in *Physcomitrella*) (De Bortoli et al. 2016).

The GLRs are mainly localized to the plasma membrane (PM) where they are thought to initiate the flux of Ca^{2+}. They are thought to be the primary drivers which modulate the membrane potential to activate the voltage-gated ion channels in the PM (Wudick et al. 2018a). Recent reports have shown localization of GLRs to vacuole (GLR2.1) (Wudick et al. 2018b), sperm cells (GLR3.3) (Wudick et al. 2018b), chloroplast (GLR3.4 and GLR3.5.2) (Teardo et al. 2015) and mitochondria (GLR3.5.1) (Teardo et al. 2015). A recent report shows that cornichon homolog (CNIH) proteins play a crucial role in GLR localization and channel activity (Wudick et al. 2018b). The fact that GLRs are located in some of the important Ca^{2+} stores in plant cell and that they may have the propensity to modulate membrane voltages makes them ideal to stimulate long-distance signal transport in plants through Ca^{2+} fluxes. We talk about this function of plants in the final chapter of the book.

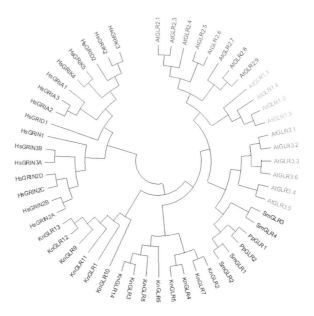

Phylogram of iGluR and GLR

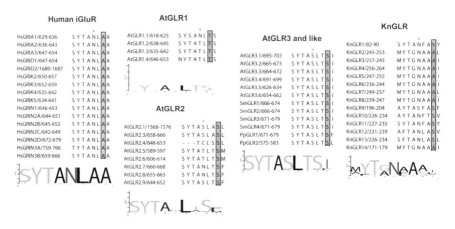

Fig. 6.2 Phylogenetic relationship of GLR. Neighbor-Joining tree showing rooted relationship to the human GLRs (marked in red). *Arabidopsis thaliana* GLRs form three distinct clades, GLR1 (in green), GLR2 (in violet) and GLR3 with others (*Physcomitrella patens* and *Selaginella moellendorfii*) GLRs (in orange). The *Klebsormidium nitens* GLR form entirely different group indicating these may be different. This fact is further emphasized when we analyzed the sequences responsible for gating. Conserved Ala was found in human but missing in most of the plant GLRs. The *Klebsormidium* shows a mix of Ala and Ser. The human, Arabidopsis and *Physcomitrella* sequences are from (Wudick et al. 2018a). MSA performed using MEGA10 and viewed using Jalview. The consensus was viewed using WebLogo

Besides this GLRs have been implicated in controlling many aspects of plant related to ABA signaling (like the synthesis of the hormone itself, stomatal dynamics and water loss) (Kang et al. 2004; Cho et al. 2009; Lu et al. 2014; Kong et al. 2015). It can regulate plant development (pollen tube growth, self-incompatibility, developmental features in moss) (Michard et al. 2011; Iwano et al. 2015; Ortiz-Ramírez et al. 2017; Wudick et al. 2018b), root biology (gravitropism, root development and lateral root initiation) (Miller et al. 2010; Vincill et al. 2013; Singh et al. 2016; Wudick et al. 2018a). Besides, newer studies have indicated its major role during plant aphid interaction and how plant perceive wounding (Vincent et al. 2017; Toyota et al. 2018).

Conclusion

GLR serves as a major Ca^{2+} permeable channel in plants. The increase in their number in vascular plants outlines their importance in plants. Also, recent years have established GLR playing a crucial role in plant physiology. The lack of structural data is one of the shortcomings of plant GLR research, but hopefully, this scenario will change with research groups actively working in this aspect as well.

References

F.C. Acher, H.-O. Bertrand, Amino acid recognition by Venus flytrap domains is encoded in an 8-residue motif. Pept. Sci. **80**, 357–366 (2005)

A. Alfieri, F.G. Doccula, R. Pederzoli, M. Grenzi, M.C. Bonza, L. Luoni, A. Candeo, N. Romano Armada, A. Barbiroli, G. Valentini, T.R. Schneider, A. Bassi, M. Bolognesi, M. Nardini, A. Costa, The structural bases for agonist diversity in an Arabidopsis thaliana glutamate receptor-like channel. Proc. Natl. Acad. Sci. U. S. A. **117**, 752–760 (2020)

J.C. Chiu, E.D. Brenner, R. DeSalle, M.N. Nitabach, T.C. Holmes, G.M. Coruzzi, Phylogenetic and expression analysis of the glutamate-receptor-like gene family in Arabidopsis thaliana. Mol. Biol. Evol. **19**, 1066–1082 (2002)

D. Cho, S.A. Kim, Y. Murata, S. Lee, S.K. Jae, H.G. Nam, J.M. Kwak, De-regulated expression of the plant glutamate receptor homolog AtGLR3.1 impairs long-term Ca2+−programmed stomatal closure. Plant J. **58**, 437–449 (2009)

S. De Bortoli, E. Teardo, I. Szabò, T. Morosinotto, A. Alboresi, Evolutionary insight into the ionotropic glutamate receptor superfamily of photosynthetic organisms. Biophys. Chem. **218**, 14–26 (2016)

M. Iwano, K. Ito, S. Fujii, M. Kakita, H. Asano-Shimosato, M. Igarashi, P. Kaothien-Nakayama, T. Entani, A. Kanatani, M. Takehisa, M. Tanaka, K. Komatsu, H. Shiba, T. Nagai, A. Miyawaki, A. Isogai, S. Takayama, Calcium signalling mediates self-incompatibility response in the Brassicaceae. Nat Plants **1**, 15128 (2015)

J. Kang, S. Mehta, F.J. Turano, The putative glutamate receptor 1.1 (AtGLR1.1) in Arabidopsis thaliana regulates abscisic acid biosynthesis and signaling to control development and water loss. Plant Cell Physiol. **45**, 1380–1389 (2004)

S.A. Kim, J.M. Kwak, S.K. Jae, M.H. Wang, H.G. Nam, Overexpression of the AtGluR2 gene encoding an Arabidopsis homolog of mammalian glutamate receptors impairs calcium utilization and sensitivity to ionic stress in transgenic plants. Plant Cell Physiol. **42**, 74–84 (2001)

D. Kong, C. Ju, A. Parihar, S. Kim, D. Cho, J.M. Kwak, Arabidopsis glutamate receptor Homolog3.5 modulates cytosolic Ca2+ level to counteract effect of Abscisic acid in seed germination. Plant Physiol. **167**, 1630–1642 (2015)

B. Lacombe, D. Becker, R. Hedrich, R. DeSalle, M. Hollmann, J.M. Kwak, J.I. Schroeder, N. Le Novère, H.G. Nam, E.P. Spalding, M. Tester, F.J. Turano, J. Chiu, G. Coruzzi, The identity of plant glutamate receptors. Science **292**, 1486–1487 (2001)

H.M. Lam, J. Chiu, M.H. Hsieh, L. Meisel, I.C. Oliveira, M. Shin, G. Coruzzi, Glutamate-receptor genes in plants. Nature **396**, 125–126 (1998)

G. Lu, X. Wang, J. Liu, K. Yu, Y. Gao, H. Liu, C. Wang, W. Wang, G. Wang, M. Liu, G. Mao, B. Li, J. Qin, M. Xia, J. Zhou, S. Jiang, H. Mo, J. Cui, N. Nagasawa, S. Sivasankar, M.C. Albertsen, H. Sakai, B.J. Mazur, M.W. Lassner, R.M. Broglie, Application of T-DNA activation tagging to identify glutamate receptor-like genes that enhance drought tolerance in plants. Plant Cell Rep. **33**, 617–631 (2014)

E. Michard, P.T. Lima, F. Borges, A.C. Silva, M.T. Portes, J.E. Carvalho, M. Gilliham, L.H. Liu, G. Obermeyer, J.A. Feijo, Glutamate receptor-like genes form Ca2+ channels in pollen tubes and are regulated by pistil D-serine. Science **332**, 434–437 (2011)

N.D. Miller, T.L. Durham Brooks, A.H. Assadi, E.P. Spalding, Detection of a Gravitropism phenotype in glutamate receptor-like 3.3 mutants of Arabidopsis thaliana using machine vision and computation. Genetics **186**, 585–593 (2010)

C. Ortiz-Ramírez, E. Michard, A.A. Simon, D.S.C. Damineli, M. Hernández-Coronado, J.D. Becker, J.A. Feijó, GLUTAMATE RECEPTOR-LIKE channels are essential for chemotaxis and reproduction in mosses. Nature **549**, 91–95 (2017)

M.B. Price, J. Jelesko, S. Okumoto, Glutamate receptor homologs in plants: Functions and evolutionary origins. Front. Plant Sci. **3**, 235 (2012)

M.B. Price, D. Kong, S. Okumoto, Inter-subunit interactions between glutamate-like receptors in Arabidopsis. Plant Signal. Behav. **8**, e27034 (2013)

S.K. Singh, C.T. Chien, I.F. Chang, The Arabidopsis glutamate receptor-like gene GLR3.6 controls root development by repressing the Kip-related protein gene KRP4. J. Exp. Bot. **67**, 1853–1869 (2016)

A.I. Sobolevsky, Structure and gating of tetrameric glutamate receptors. J. Physiol. **593**, 29–38 (2015)

N.R. Stephens, Z. Qi, E.P. Spalding, Glutamate receptor subtypes evidenced by differences in desensitization and dependence on the GLR3.3 and GLR3.4 Genes. Plant Physiol. **146**, 529–538 (2008)

D. Tapken, M. Hollmann, Arabidopsis thaliana glutamate receptor ion channel function demonstrated by ion pore transplantation. J. Mol. Biol. **383**, 36–48 (2008)

D. Tapken, U. Anschütz, L.H. Liu, T. Huelsken, G. Seebohm, D. Becker, M. Hollmann, A plant homolog of animal glutamate receptors is an ion channel gated by multiple hydrophobic amino acids. Sci. Signal. **6**, ra47 (2013)

E. Teardo, L. Carraretto, S. De Bortoli, A. Costa, S. Behera, R. Wagner, F. Lo Schiavo, E. Formentin, I. Szabo, Alternative Splicing-Mediated targeting of the Arabidopsis GLUTAMATE RECEPTOR3.5 to Mitochondria affects Organelle Morphology1. Plant Physiol **167**, 216–227 (2015)

M. Toyota, D. Spencer, S. Sawai-Toyota, W. Jiaqi, T. Zhang, A.J. Koo, G.A. Howe, S. Gilroy, Glutamate triggers long-distance, calcium-based plant defense signaling. Science **361**, 1112–1115 (2018)

S.F. Traynelis, L.P. Wollmuth, C.J. McBain, F.S. Menniti, K.M. Vance, K.K. Ogden, K.B. Hansen, H. Yuan, S.J. Myers, R. Dingledine, Glutamate receptor ion channels: Structure, regulation, and function. Pharmacol. Rev. **62**, 405–496 (2010)

F.J. Turano, G.R. Panta, M.W. Allard, P. van Berkum, The putative glutamate receptors from plants are related to two Superfamilies of animal Neurotransmitter receptors via distinct evolutionary mechanisms. Mol. Biol. Evol. **18**, 1417–1420 (2001)

F.J. Turano, M.J. Muhitch, F.C. Felker, M.B. McMahon, The putative glutamate receptor 3.2 from Arabidopsis thaliana (AtGLR3.2) is an integral membrane peptide that accumulates in rapidly growing tissues and persists in vascular-associated tissues. Plant Sci. **163**, 43–51 (2002)

E. Twomey, A.I. Sobolevsky, Structural mechanisms of gating in ionotropic glutamate receptors. Biochemistry **57**, 267–276 (2018)

T.R. Vincent, M. Avramova, J. Canham, P. Higgins, N. Bilkey, S.T. Mugford, M. Pitino, M. Toyota, S. Gilroy, A.J. Miller, S.A. Hogenhout, D. Sanders, Interplay of plasma membrane and vacuolar ion channels, together with BAK1, elicits rapid cytosolic calcium elevations in Arabidopsis during aphid feeding. Plant Cell **29**, 1460–1479 (2017)

E.D. Vincill, A.M. Bieck, E.P. Spalding, Ca2+ conduction by an amino acid-gated Ion channel related to glutamate Receptors1. Plant Physiol. **159**, 40–46 (2012)

E.D. Vincill, A.E. Clarin, J.N. Molenda, E.P. Spalding, Interacting glutamate receptor-like proteins in phloem regulate lateral root initiation in Arabidopsis. Plant Cell **25**, 1304–1313 (2013)

M. Weiland, S. Mancuso, F. Baluska, Signalling via glutamate and GLRs in Arabidopsis thaliana. Funct. Plant Biol. **43**, 1–25 (2016)

M.M. Wudick, E. Michard, C. Oliveira Nunes, J.A. Feijó, Comparing plant and animal glutamate receptors: Common traits but different fates? J. Exp. Bot. **69**, 4151–4163 (2018a)

M.M. Wudick, M.T. Portes, E. Michard, P. Rosas-Santiago, M.A. Lizzio, C.O. Nunes, C. Campos, D. Santa Cruz Damineli, J.C. Carvalho, P.T. Lima, O. Pantoja, J.A. Feijó, CORNICHON sorting and regulation of GLR channels underlie pollen tube Ca(2+) homeostasis. Science **360**, 533–536 (2018b)

S. Zhu, E. Gouaux, Structure and symmetry inform gating principles of ionotropic glutamate receptors. Neuropharmacology **112**, 11–15 (2017)

Chapter 7
Plant Ligand-Gated Channels 2: CNGC

Contents

Introduction

The classification of cyclic nucleotide-gated channels (CNGC, and henceforth in the chapter CNGC will imply plant-specific channels) as only ligand-gated channels is incorrect. The CNGCs, similarly pose another challenging question, whether they are activated by cyclic nucleotides in the plant system? The cyclic nucleotide (CN) (namely adenosine-3′, 5′-cyclic monophosphate (cAMP) and guanosine-3′, 5′-cyclic monophosphate (cGMP)) ligands modulate an array of proteins (including the CNGCs) (Gehring and Turek 2017; Świeżawska et al. 2018; Duszyn et al. 2019). The CN based signaling system is very well worked out in the animal system. But in plants, this topic remained controversial, again due to the inability to pinpoint the molecular identity of the nucleotide cyclase (NC) and phosphodiesterase (PDE) (both enzymes involved in the regulation of CN in the cell) (Duszyn et al. 2019). However, the presence of cyclic nucleotide was proved in plants indicating that it can regulate CNGC (and other CN regulated proteins). Lately, there have been some reports that some proteins with modified domains can produce CNs in plants. Later searches have proved that plant NCs are not canonical (compared to the animal counterparts), and have evolved with significant differences in their sequences (Gehring and Turek 2017). Several NCs have been identified in plants with modified domains (Gehring and Turek 2017). The existence of PDEs in plants was initially proven through analyzing protein activity in plant extracts (Duszyn et al. 2019). Till date, there has been only one report from *Marchantia polymorpha* that has a protein with both nucleotide cyclase and PDE (Kasahara et al. 2016). However, similar proteins have not been identified in other plants (Duszyn et al. 2019). But there are

G. K. Pandey, S. K. Sanyal, *Functional Dissection of Calcium Homeostasis and Transport Machinery in Plants*, SpringerBriefs in Plant Science,
https://doi.org/10.1007/978-3-030-58502-0_7

several reports on the role of CN in regulating plant physiology (Gehring and Turek 2017). The current understanding indicates that the plant CNGCs, like their animal counterparts, are regulated by CNs. However, recent report indicate that CN may be co-factors rather than activation triggers for CNGC (Dietrich et al. 2020)

As we discuss our initial statement, regarding the classification of CNGCs- if they are voltage-gated or ligand-gated? The animal system has two sets of ion channels that are regulated by CNs- the cyclic-nucleotide gated (CNG) and hyperpolarization-activated cyclic nucleotide–modulated (HCN) channels (Craven and Zagotta 2006; Cukkemane et al. 2011). The CNG mostly rely on CNs for their regulation and the HCNs rely mostly on the membrane polarization (i.e., they are voltage-gated) and CNs (Craven and Zagotta 2006). The CNGCs have been proven to be modulated by voltages (discussed little later in the chapter) and hence it would be incorrect to classify them as solely ligand-gated. But, in this writeup, we have classified them as ligand-gated since plant research community majorly classifies these channels as ligand-gated.

Structural Organization and Interpretation of Function

The plant CNGCs have a problem similar to the GLRs, there are no reports on their crystal structure. As such the information on them is often derived from sequence comparison or homology modelling of proteins (based on other available CNG structures). The CNGCs share a similar structure to an isolated domain of TPC1 (i.e., it has a single shaker like domain) with some modifications. The core structure of 6 transmembrane (TM) helices with a pore region to conduct ions (non-selective cation channel) stays the same (Talke et al. 2003). The differences occur at the N-terminal and after the 6th TM helix, where CNGC modulating regions are present. The CNGCs are also modulated by Calmodulin (CaM) protein, and recent analysis has shown that some CNGCs can have as many as 4 CaM binding sites in the N-terminal region, the cyclic nucleotide-binding domain (CNBD), the IQ motif and the C-terminal (DeFalco et al. 2016; Fischer et al. 2017). The CNBD also is the CN binding region of CNGC connected to the 6th TM helix with a linker. The structure of CNGC is indicated in Fig. 7.1. In the membrane, the polypeptides tetramerize (whether they form homomers or heteromers, we discuss this little later) to form a functional channel with the N- and C- terminal towards the cytosol. Basically there are two hypothesis regarding CNGC channel modulation – (1) CN binding to CNBD would lead to the opening of the "gate" to allow ions into the cytoplasm and (2) Ca^{2+} bound CaM regulates the CNGC channel (either by opening or closing it) (Dietrich et al. 2020). We now look at the structure of CNGC with more detail to understand these functions of CNGC.

CNGC12 has a site at its N-terminal where it can bind CaM and this can regulates the CNGC12's function (DeFalco et al. 2016). CNGC12 has a role in plant immunity and the binding of CaM to its N-terminal inhibits CNGC12 mediated auto-immunity activation in plants (DeFalco et al. 2016). The animals CNG also

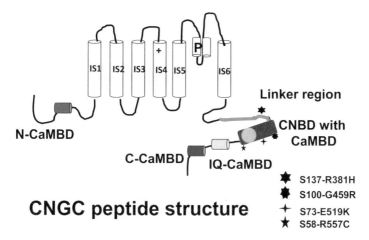

IS1 IS2 IS3 IS4 IS5 IS6

Linker region

N-CaMBD

CNBD with CaMBD

C-CaMBD **IQ-CaMBD**

S137-R381H
S100-G459R
S73-E519K
S58-R557C

CNGC peptide structure

Fig. 7.1 Structure of plant CNGC. The domain structure of CNGC is presented. The N-terminal CaM binding domain (N-CaMBD) as reported by DeFalco et al. (2016). The core structure is similar to a shaker channel domain. The 6th helix (S6) is linked to CNBD with a linker. The cyclic nucleotide-binding domain (CNBD) has the PBC cassette, hinge region and a CaMBD. The CNBD is followed by IQ motif (another CaMBD) and the final C-terminal CaMBD (C-CaMBD). The suppressor of chimera CNGC11/12 (*cpr22*) mutations are marked (locations according to Abdel-Hamid et al. (2013). Detailed description in the text

have similar N-terminal CaM binding motifs (known as LQ site) and plays a role in channel regulation (Ungerer et al. 2011). If this site is functional in other Arabidopsis CNGCs is unknown at the moment.

The shaker like domain of CNGC, as already mentioned, is similar to TPC1's shaker like domain and hence is not discussed here. The S4 TM helix of CNGC, however, can act as a voltage sensor even though they have fewer charged residue (Arg or Lys) (Duszyn et al. 2019). Several studies have proved that CNGCs can respond to voltage-gating (heterologous and *in planta* mutant analysis) (Leng et al. 1999, 2002; Hua et al. 2003; Ali et al. 2007; Gao et al. 2012, 2016; Wang et al. 2013; Mori et al. 2018). So, this indicates their similarity to the HCN rather than CNG channels. However, CNGC18 works as a voltage-independent CN gated channel, providing shreds of evidence of a mixed regulation (Demidchik and Shabala 2018; Demidchik et al. 2018). Next, we move to the pore sequence. The 20 member Arabidopsis CNGC family has six different selectivity filter (for transporting ion selection) indicated in Fig. 7.2 (Jammes et al. 2011). We have already mentioned the non-selective nature of CNGCs pore (it can permeate K^+ and Na^+, and recent studies have indicated that GQN, GQG and GQS triplets are Ca^{2+} selective (Leng et al. 1999, 2002; Christopher et al. 2007; Demidchik et al. 2018; Mori et al. 2018). The CNG channels have a different filter (GET) allowing K^+, Na^+, Ca^{2+} and Mg^{2+} transport (the latter two are transported at a comparatively slower rate). The HCN carry the typical GYG motif for K^+ conductance (they also mediate Na^+ and Ca^{2+}) (Craven and Zagotta 2006; Cukkemane et al. 2011).

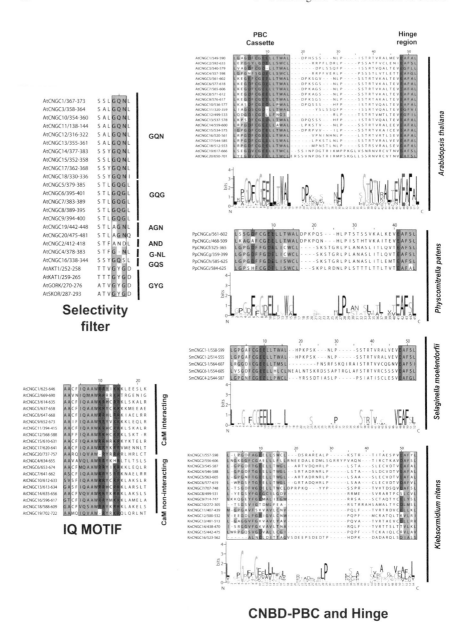

Fig. 7.2 Multiple sequence alignment of important regions of CNGC. The selectivity filter is shown in the figure. The selectivity filter of CNBD containing K⁺ channels (AKT1, KAT1, GORK and SKOR) are shown for comparison, The IQ motif sequence is shown with a division of CaM interacting and CaM non-interacting CNGCs according to Fischer et al. (2017). The PBC cassette and hinge region of the CNBD domain of CNGC. The consensus sequence varies with species. MSA performed by MEGA10 and viewed using Jalview. The consensus was viewed using WebLogo

Subsequently, we move to the main regulatory segment in the CNGC channel, the portion beyond the 6th TM helix. From the animal channels, it is known that the C-linker region performs dual functions- (1) helps in tetramerization of the subunits to form a channel, and (2) it relays the CN binding to the channel gate (Cukkemane et al. 2011). Similar information is lacking for CNGC, except for mutation analysis. A suppressor analysis to investigate genes involved in plant defense response resulted in the generation of a chimeric CNGC11/12 (N-terminal of CNGC11 and C-terminal of CNGC12), which constitutively activated programmed cell death (PCD) (Yoshioka et al. 2006). Several mutants to this chimera were generated that reversed the phenotype expressed by this chimera (important ones are marked in Fig. 7.1). These mutants have increased our understanding of CNGC functioning. The first of them to be characterized was Glu519Lys (E519K) (labelled S73), which resulted in a non-functional channel (Baxter et al. 2008). It was hypothesized that this residue is important for the interaction between C-linker and CNBD. Another two mutations Arg381His (R381H) in the C-linker (labelled S137) and Gly459Arg (G459R) in the CNDB (labelled S100) provided the first indication that the CNGCs may form tetramers to function as a channel (Abdel-Hamid et al. 2013). The information regarding the nature of tetramers of CNGC channel is still not clear, with the yeast complementation experiments indicating that homo-tetramers of CNGC can form a functional channel (Abdel-Hamid et al. 2013). Hetero-tetramerization has been also reported for CNGC8-CNGC18 and CNGC2-CNGC4 complex (Pan et al. 2019; Tian et al. 2019).

The next segment sets the major difference between the CNGC and the CNG (also HCN) channels. The CNGC CNBD has the CN binding domain overlapping a CaM binding domain (CaMBD) (with a signature motif WRTW, CaMBD is not present in CNGC20) (Fischer et al. 2013). However, the CNG and HCN have this domain separately (Duszyn et al. 2019). The CNBD has 3 α-helix (αA, αB and αC) and 10 β-sheets, and the CaMBD is predicted to be present at the αC helix (Fischer et al. 2017). The CNBD itself has two very important regions that help in the ligand binding and are thought to be the signature motif for CNGC identification (-the phosphate binding cassette (PBC) and hinge region, both regions before the CaMBD in the CNGC) (Duszyn et al. 2019). PBC binds to the CNs and hinge region helps in ligand binding selectivity and efficacy (Zelman et al. 2012). The PBC cassette varies among the CNGCs (from different species, implying its flexibility) and it is highlighted in Fig. 7.2 (Saand et al. 2015). The last of the important mutations Arg557Cys (R557C) labelled S58 and S136 (a different mutation, Glu543 to a premature stop codon obliterating everything other than a partial αB, upstream of αB intact) disrupted the CaM binding site within the CNBD showing the importance of CaM in the regulation of CNGC channel (Chin et al. 2010). The follow-up studies have indicated that CNGCs may have multiple CaM binding sites (DeFalco et al. 2016). The IQ motif (more universal in the CNGCs, but does not allow CNGC-CaM interaction for all) that is the second CaM binding domain in CNGCs (depicted in Fig. 7.2) (DeFalco et al. 2016; Fischer et al. 2017). This is followed by another site in the C-terminal end (proved in CNGC12) (DeFalco et al. 2016). These multiple sites have given rise to an important hypothesis regarding CaM-CNGC regulation.

The CaM can be kept bound in the C-terminal of the CNGC even without Ca^{2+} to serve as embedded Ca^{2+} sensor (this happens in the animal system) (DeFalco et al. 2016). Binding of CN to the CNBD makes conformational changes to the gate of the CNGC channel that allow the opening and Ca^{2+} entry in the cytosol. When there is excess Ca^{2+} in the cytosol, CaM can sense it (due to Ca^{2+} binding in its EF-hands) and make further conformational changes to close the gate (Fischer et al. 2017). This is shown in Fig. 7.3. The CaM can also compete with CNs due to the presence of a CaMBD in CNBD and causes inactivation of the channel (not shown in the Fig. 7.3). Further, the CNGC8-CNGC18 heteromer needs CaM for its activation, and elevation of cytosolic Ca^{2+} causes dissociation of Ca^{2+}-CaM inactivating the heteromer (Pan et al. 2019; Dietrich et al. 2020). Additionally, phosphorylation plays an important role in modulating channel properties (Curran et al. 2011; Zhou et al. 2014; He et al. 2019; Tian et al. 2019; Wang et al. 2019; Yu et al. 2019). Hence, the mode of CNGC regulation is more dynamic and it may be possible that one model may not fit all functional attributes.

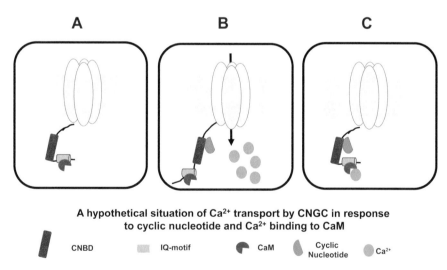

A hypothetical situation of Ca^{2+} transport by CNGC in response to cyclic nucleotide and Ca^{2+} binding to CaM

CNBD   IQ-motif   CaM   Cyclic Nucleotide Ca^{2+}

Fig. 7.3 A hypothetical model of regulation of CNGC channel. (**a**) A CNGC homo-tetramer is shown which is inactive in normal state. CaM is bound to the IQ-motif in this state and the gate is closed. (**b**) In response to CN binding to CNBD, there is a conformational change and the gate opens allowing Ca^{2+} into the cytosol. (**c**) The Ca^{2+} binding to CaM causes structural changes that causes closing of the gate to stop Ca^{2+} inflow into the cytoplasm. These situations are not all-encompassing in case of CNGC regulation. Other possibilities are also feasible for CNGC regulation. This is adopted and modified from Fischer et al. (2017)

Evolution, Localization and Biological Function of CNGC

The CNGC share their domain characteristics with CN binding K$^+$ channels (like AKT1, KAT1, GORK and SKOR channels) by having a similar shaker like structure and CNBD domain (Talke et al. 2003). Though the selectivity filter sets it apart, yet the evolutionary origin of CNGC remains unclear (Saand et al. 2015). The animal (10 including CNG and HCN) and plant CNGCs (20 in Arabidopsis) vary in their number (Maser et al. 2001; Craven and Zagotta 2006). The CNGCs have been identified in several plant species, algae and prokaryotes. The prokaryotic CNG channel has a difference in its structure compared to eukaryotic CNG (and CNGC). Figure 7.4 depicts a phylogenetic tree of the CNGC. We have already mentioned

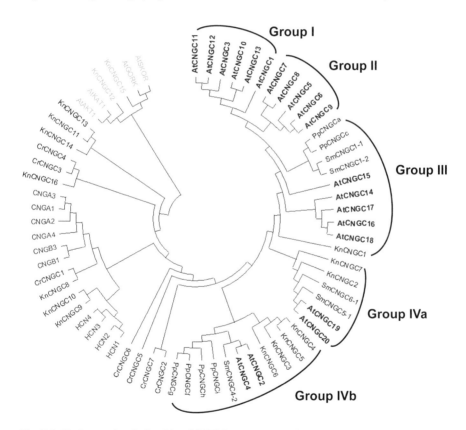

Fig. 7.4 Phylogenetic relationship of CNGC. Neighbor-Joining tree show the rooted link to the *Arabidopsis thaliana* K$^+$ channels with CNBD (marked in green). The human CNG and HCN form a separate group (marked in red). The classification of CNGC channels into clades is according to Maser et al. (2001). The *Chlamydomonas reinharditii* (Cr), *Physcomitrella patens* (Pp), and *Selaginella moellendorfii* (Sm) CNGCs sequence were taken from Saand et al. (2015). *Arabidopsis thaliana* (bold) and human sequences are from Demidchik et al. (2018). The *Klebsormidium nitens* (Kn) were identified using the Arabidopsis CNGC in the BLAST search. Tree generated using MEGA10

that the PBC region is highly variable and a single consensus motif cannot cover the full diversity of this region (Saand et al. 2015). The algal CNGCs do not have plant-specific PBC and hinge sequence and the fungal CNGCs lack CNBD and the plant CNBD has the overlapping CaMBD (Saand et al. 2015). All these fact points that plant adopted the CNGC according to their needs and hence it developed unique properties relating to its modulation and ion transport.

The presence of TM helices dictates that the CNGC will be embedded into membrane and research has proven their localization in various compartments modulating Ca^{2+} entry into the cytoplasm (and nucleus in special case). Some CNGCs display dual localization- they can be present in two different compartments (interpretation based on the different localization predictions). The plasma membrane harbors CNGC3, CNGC5, CNGC6, CNGC10, CNGC11, CNGC12 and CNGC18 and tonoplast has CNGC7, CNGC8, CNGC19 and CNGC20 (Duszyn et al. 2019). CNGC10 (ER, Golgi cisternae and vesicles), CNGC18 (vesicles) and CNGC7 (plasma membrane) show shreds of evidence of dual localization (Duszyn et al. 2019). The *Medicago truncatula* CNGCs are localized to the nucleus (they have been discussed in Chap. 3).

The role of CNGC have been extensively worked in its relationship to biotic stress (Clough et al. 2000; Balagué et al. 2003; Jurkowski et al. 2004; Yoshioka et al. 2006; Moeder et al. 2011; Meena et al. 2019; Tian et al. 2020). We have extensively talked about CNGC in relation to symbiosis in Chaps. 3 and 4. Additionally, there is extensive proof of CNGCs functioning during pollen tube development, root hair growth and cell death (Frietsch et al. 2007; Tunc-Ozdemir et al. 2013; Zhang et al. 2017; Brost et al. 2019; Cheung 2019; Yu et al. 2019). Besides the obvious role in plant development, CNGCs play an active role in sensing plant abiotic stress as well (Jha et al. 2016; Duszyn et al. 2019).

Conclusion

The ligand-gated (and voltage-gated) CNGC play an important role in Ca^{2+} transport in the cell. Their regulation by NCs will require further research, and the regulation of the channel (a homomer or heteromer) by CaM and phosphorylation presents a complex picture. But there is no doubt regarding the plethora of important physiological processes regulated by CNGC in plants.

References

H. Abdel-Hamid, K. Chin, W. Moeder, D. Shahinas, D. Gupta, K. Yoshioka, A suppressor screen of the chimeric AtCNGC11/12 reveals residues important for intersubunit interactions of cyclic nucleotide-gated ion Channels. Plant Physiol. **162**, 1681–1693 (2013)

R. Ali, W. Ma, F. Lemtiri-Chlieh, D. Tsaltas, Q. Leng, S. von Bodman, G.A. Berkowitz, Death don't have no mercy and neither does calcium: Arabidopsis CYCLIC NUCLEOTIDE GATED CHANNEL2 and innate immunity. Plant Cell **19**, 1081–1095 (2007)

C. Balagué, B. Lin, C. Alcon, G. Flottes, S. Malmström, C. Köhler, G. Neuhaus, G. Pelletier, F. Gaymard, D. Roby, HLM1, an essential signaling component in the hypersensitive response, is a member of the cyclic nucleotide–Gated Channel Ion Channel Family. Plant Cell **15**, 365–379 (2003)

J. Baxter, W. Moeder, W. Urquhart, D. Shahinas, K. Chin, D. Christendat, H.G. Kang, M. Angelova, N. Kato, K. Yoshioka, Identification of a functionally essential amino acid for Arabidopsis cyclic nucleotide gated ion channels using the chimeric AtCNGC11/12 gene. Plant J. **56**, 457–469 (2008)

C. Brost, T. Studtrucker, R. Reimann, P. Denninger, J. Czekalla, M. Krebs, B. Fabry, K. Schumacher, G. Grossmann, P. Dietrich, Multiple cyclic nucleotide-gated channels coordinate calcium oscillations and polar growth of root hairs. Plant J. **99**, 910–923 (2019)

A.Y. Cheung, Plant biology: To live, or not to live, that is the question. Curr. Biol. **29**, R1186–r1189 (2019)

K. Chin, W. Moeder, H. Abdel-Hamid, D. Shahinas, D. Gupta, K. Yoshioka, Importance of the αC-helix in the cyclic nucleotide binding domain for the stable channel regulation and function of cyclic nucleotide gated ion channels in Arabidopsis. J. Exp. Bot. **61**, 2383–2393 (2010)

D.A. Christopher, T. Borsics, C.Y. Yuen, W. Ullmer, C. Andème-Ondzighi, M.A. Andres, B.H. Kang, L.A. Staehelin, The cyclic nucleotide gated cation channel AtCNGC10 traffics from the ER via Golgi vesicles to the plasma membrane of Arabidopsis root and leaf cells. BMC Plant Biol. **7**, 48 (2007)

S.J. Clough, K.A. Fengler, I. Yu, B. Lippok, R.K. Smith, A.F. Bent, The Arabidopsis dnd1 "defense, no death" gene encodes a mutated cyclic nucleotide-gated ion channel. Proc. Natl. Acad. Sci. U. S. A. **97**, 9323–9328 (2000)

K.B. Craven, W.N. Zagotta, CNG and HCN channels: Two peas, one pod. Annu. Rev. Physiol. **68**, 375–401 (2006)

A. Cukkemane, R. Seifert, U.B. Kaupp, Cooperative and uncooperative cyclic-nucleotide-gated ion channels. Trends Biochem. Sci. **36**, 55–64 (2011)

A. Curran, I.F. Chang, C.L. Chang, S. Garg, R.M. Miguel, Y.D. Barron, Y. Li, S. Romanowsky, J.C. Cushman, M. Gribskov, A.C. Harmon, J.F. Harper, Calcium-dependent protein kinases from Arabidopsis show substrate specificity differences in an analysis of 103 substrates. Front. Plant Sci. **2**, 36 (2011)

T.A. DeFalco, C.B. Marshall, K. Munro, H.G. Kang, W. Moeder, M. Ikura, W.A. Snedden, K. Yoshioka, Multiple Calmodulin-binding sites positively and negatively regulate Arabidopsis CYCLIC NUCLEOTIDE-GATED CHANNEL12. Plant Cell **28**, 1738–1751 (2016)

V. Demidchik, S. Shabala, Mechanisms of cytosolic calcium elevation in plants: The role of ion channels, calcium extrusion systems and NADPH oxidase-mediated 'ROS-Ca(2+) Hub'. Funct. Plant Biol. **45**, 9–27 (2018)

V. Demidchik, S. Shabala, S. Isayenkov, T.A. Cuin, I. Pottosin, Calcium transport across plant membranes: Mechanisms and functions. New Phytol. **220**, 49–69 (2018)

P. Dietrich, W. Moeder, K. Yoshioka, Plant cyclic nucleotide-gated channels: New insights on their functions and regulation. Plant Physiol. **184**, 27–38 (2020)

M. Duszyn, B. Świeżawska, A. Szmidt-Jaworska, K. Jaworski, Cyclic nucleotide gated channels (CNGCs) in plant signalling-current knowledge and perspectives. J. Plant Physiol. **241**, 153035 (2019)

C. Fischer, A. Kugler, S. Hoth, P. Dietrich, An IQ domain mediates the interaction with Calmodulin in a plant cyclic nucleotide-Gated Channel. Plant Cell Physiol. **54**, 573–584 (2013)

C. Fischer, T.A. DeFalco, P. Karia, W.A. Snedden, W. Moeder, K. Yoshioka, P. Dietrich, Calmodulin as a Ca2+-sensing subunit of Arabidopsis Cyclic Nucleotide-Gated Channel complexes. Plant Cell Physiol. **58**, 1208–1221 (2017)

S. Frietsch, Y.F. Wang, C. Sladek, L.R. Poulsen, S.M. Romanowsky, J.I. Schroeder, J.F. Harper, A cyclic nucleotide-gated channel is essential for polarized tip growth of pollen. Proc. Natl. Acad. Sci. U. S. A. **104**, 14531–14536 (2007)

F. Gao, X. Han, J. Wu, S. Zheng, Z. Shang, D. Sun, R. Zhou, B. Li, A heat-activated calcium-permeable channel--Arabidopsis cyclic nucleotide-gated ion channel 6–is involved in heat shock responses. Plant J. **70**, 1056–1069 (2012)

Q.F. Gao, L.L. Gu, H.Q. Wang, C.F. Fei, X. Fang, J. Hussain, S.J. Sun, J.Y. Dong, H. Liu, Y.F. Wang, Cyclic nucleotide-gated channel 18 is an essential Ca2+ channel in pollen tube tips for pollen tube guidance to ovules in Arabidopsis. Proc. Natl. Acad. Sci. U. S. A. **113**, 3096–3101 (2016)

C. Gehring, I.S. Turek, Cyclic nucleotide monophosphates and their Cyclases in plant signaling. Front. Plant Sci. **8**, 1704 (2017)

Y. He, J. Zhou, X. Meng, Phosphoregulation of Ca(2+) influx in plant immunity. Trends Plant Sci. **24**, 1067–1069 (2019)

B.G. Hua, R.W. Mercier, Q. Leng, G.A. Berkowitz, Plants do it differently. A new basis for potassium/sodium selectivity in the pore of an Ion Channel. Plant Physiol. **132**, 1353–1361 (2003)

F. Jammes, H.C. Hu, F. Villiers, R. Bouten, J.M. Kwak, Calcium-permeable channels in plant cells. FEBS J. **278**, 4262–4276 (2011)

S.K. Jha, M. Sharma, G.K. Pandey, Role of cyclic nucleotide gated channels in stress Management in Plants. Curr. Genomics **17**, 315–329 (2016)

G.I. Jurkowski, R.K. Smith Jr., I.C. Yu, J.H. Ham, S.B. Sharma, D.F. Klessig, K.A. Fengler, A.F. Bent, Arabidopsis DND2, a second cyclic nucleotide-gated ion channel gene for which mutation causes the "defense, no death" phenotype. Mol. Plant-Microbe Interact. **17**, 511–520 (2004)

M. Kasahara, N. Suetsugu, Y. Urano, C. Yamamoto, M. Ohmori, Y. Takada, S. Okuda, T. Nishiyama, H. Sakayama, T. Kohchi, F. Takahashi, An adenylyl cyclase with a phosphodiesterase domain in basal plants with a motile sperm system. Sci. Rep. **6**, 39232 (2016)

Q. Leng, R.W. Mercier, W. Yao, G.A. Berkowitz, Cloning and first functional characterization of a plant cyclic nucleotide-gated cation channel. Plant Physiol. **121**, 753–761 (1999)

Q. Leng, R.W. Mercier, B.G. Hua, H. Fromm, G.A. Berkowitz, Electrophysiological analysis of cloned cyclic nucleotide-gated ion channels. Plant Physiol. **128**, 400–410 (2002)

P. Maser, S. Thomine, J.I. Schroeder, J.M. Ward, K. Hirschi, H. Sze, I.N. Talke, A. Amtmann, F.J. Maathuis, D. Sanders, J.F. Harper, J. Tchieu, M. Gribskov, M.W. Persans, D.E. Salt, S.A. Kim, M.L. Guerinot, Phylogenetic relationships within cation transporter families of Arabidopsis. Plant Physiol. **126**, 1646–1667 (2001)

M.K. Meena, R. Prajapati, D. Krishna, K. Divakaran, Y. Pandey, M. Reichelt, M.K. Mathew, W. Boland, A. Mithöfer, J. Vadassery, The Ca(2+) channel CNGC19 regulates Arabidopsis defense against Spodoptera herbivory. Plant Cell **31**, 1539–1562 (2019)

W. Moeder, W. Urquhart, H. Ung, K. Yoshioka, The role of cyclic nucleotide-gated ion channels in plant immunity. Mol. Plant **4**, 442–452 (2011)

I.C. Mori, Y. Nobukiyo, Y. Nakahara, M. Shibasaka, T. Furuichi, M. Katsuhara, A cyclic nucleotide-gated channel, HvCNGC2-3, is activated by the co-presence of Na+ and K+ and permeable to Na+ and K+ non-selectively. Plants (Basel) **7**, 61 (2018)

Y. Pan, X. Chai, Q. Gao, L. Zhou, S. Zhang, L. Li, S. Luan, Dynamic interactions of plant CNGC subunits and Calmodulins drive oscillatory Ca(2+) channel activities. Dev. Cell **48**, 710–725 (2019).e715

M.A. Saand, Y.P. Xu, J.P. Munyampundu, W. Li, X.R. Zhang, X.Z. Cai, Phylogeny and evolution of plant cyclic nucleotide-gated ion channel (CNGC) gene family and functional analyses of tomato CNGCs. DNA Res. **22**, 471–483 (2015)

B. Świeżawska, M. Duszyn, K. Jaworski, A. Szmidt-Jaworska, Downstream targets of cyclic nucleotides in plants. Front. Plant Sci. **9**, 1428 (2018)

I.N. Talke, D. Blaudez, F.J. Maathuis, D. Sanders, CNGCs: Prime targets of plant cyclic nucleotide signalling? Trends Plant Sci. **8**, 286–293 (2003)

W. Tian, C. Hou, Z. Ren, C. Wang, F. Zhao, D. Dahlbeck, S. Hu, L. Zhang, Q. Niu, L. Li, B.J. Staskawicz, S. Luan, A calmodulin-gated calcium channel links pathogen patterns to plant immunity. Nature **572**, 131–135 (2019)

W. Tian, C. Wang, Q. Gao, L. Li, S. Luan, Calcium spikes, waves and oscillations in plant development and biotic interactions. Nat. Plants **6**, 750–759 (2020)

M. Tunc-Ozdemir, C. Tang, M.R. Ishka, E. Brown, N.R. Groves, C.T. Myers, C. Rato, L.R. Poulsen, S. McDowell, G. Miller, R. Mittler, J.F. Harper, A cyclic nucleotide-gated channel (CNGC16) in pollen is critical for stress tolerance in pollen reproductive development. Plant Physiol. **161**, 1010–1020 (2013)

N. Ungerer, N. Mücke, J. Broecker, S. Keller, S. Frings, F. Möhrlen, Distinct binding properties distinguish LQ-type calmodulin-binding domains in cyclic nucleotide-gated channels. Biochemistry **50**, 3221–3228 (2011)

Y.F. Wang, S. Munemasa, N. Nishimura, H.M. Ren, N. Robert, M. Han, I. Puzõrjova, H. Kollist, S. Lee, I. Mori, J.I. Schroeder, Identification of cyclic GMP-activated nonselective Ca2+-permeable Cation channels and associated CNGC5 and CNGC6 genes in Arabidopsis guard cells. Plant Physiol. **163**, 578–590 (2013)

J. Wang, X. Liu, A. Zhang, Y. Ren, F. Wu, G. Wang, Y. Xu, C. Lei, S. Zhu, T. Pan, Y. Wang, H. Zhang, F. Wang, Y.Q. Tan, X. Jin, S. Luo, C. Zhou, X. Zhang, J. Liu, S. Wang, L. Meng, X. Chen, Q. Lin, X. Guo, Z. Cheng, Y. Tian, S. Liu, L. Jiang, C. Wu, E. Wang, J.M. Zhou, Y.F. Wang, H. Wang, J. Wan, A cyclic nucleotide-gated channel mediates cytoplasmic calcium elevation and disease resistance in rice. Cell Res. **29**, 820–831 (2019)

K. Yoshioka, W. Moeder, H.G. Kang, P. Kachroo, K. Masmoudi, G. Berkowitz, D.F. Klessig, The chimeric Arabidopsis CYCLIC NUCLEOTIDE-GATED ION CHANNEL11/12 activates multiple pathogen resistance responses. Plant Cell **18**, 747–763 (2006)

X. Yu, G. Xu, B. Li, L. de Souza Vespoli, H. Liu, W. Moeder, S. Chen, M.V.V. de Oliveira, S. Ariádina de Souza, W. Shao, B. Rodrigues, Y. Ma, S. Chhajed, S. Xue, G.A. Berkowitz, K. Yoshioka, P. He, L. Shan, The receptor kinases BAK1/SERK4 regulate Ca(2+) channel-mediated cellular homeostasis for cell death containment. Curr. Biol. **29**, 3778–3790 (2019). e3778

A.K. Zelman, A. Dawe, C. Gehring, G.A. Berkowitz, Evolutionary and structural perspectives of plant cyclic nucleotide-gated cation channels. Front. Plant Sci. **3**, 95 (2012)

S. Zhang, Y. Pan, W. Tian, M. Dong, H. Zhu, S. Luan, L. Li, Arabidopsis CNGC14 mediates calcium influx required for tip growth in root hairs. Mol. Plant **10**, 1004–1006 (2017)

L. Zhou, W. Lan, Y. Jiang, W. Fang, S. Luan, A calcium-dependent protein kinase interacts with and activates a calcium channel to regulate pollen tube growth. Mol. Plant **7**, 369–376 (2014)

Chapter 8
Annexin and Mechanosensitive Channel

Contents

Introduction

Ca^{2+} can be transported through ion channels, which strictly do not adhere to the basic shaker domain architecture (as we have already seen with GLR). The ANN and MSc channels represent two unique classes of Ca^{2+} transport elements in plants that have different topology compared to the ion channels discussed earlier. We are attempting to present their detail function in this chapter.

Annexins

The ANN are a unique class of proteins and have derived their name from the most important function they can perform in the cell i.e., annexing or joining two membranes (Konopka-Postupolska and Clark 2017). They perform this particular function in both animal and plant. Besides this function, they have been also implicated in allowing Ca^{2+} import into the cytosol, which is the main focus of this chapter. The basic paradigm of ANN functioning is that these proteins bind to the membrane (it can be plasma membrane (PM) or organellar membrane) mostly depending on Ca^{2+} (Laohavisit and Davies 2011; Clark et al. 2012; Konopka-Postupolska and Clark 2017). Besides annexing the membranes, it can also form pores in the membrane that allows Ca^{2+} transport. They are reported to oligomerize (Hofmann et al. 2002) and this is supposed to help them in their channel function (Clark et al. 2012). We now look into the structural detail of ANN to understand its mode of function.

G. K. Pandey, S. K. Sanyal, *Functional Dissection of Calcium Homeostasis and Transport Machinery in Plants*, SpringerBriefs in Plant Science,
https://doi.org/10.1007/978-3-030-58502-0_8

As like other proteins that are common in both animal and plants, animal ANN have been worked out in greater detail than plants. But the crystal structure of plant ANN have also been worked out, aiding our understanding of plant ANN (Hofmann et al. 2000, 2002; Hu et al. 2008). An ANN molecule has an N-terminal followed by four domains (also called annexin repeats in Fig. 8.1). These domains have Ca^{2+} binding site located at the more "convex surface"-which aids the attachment of ANN to the membrane. The N-terminal may also help ANN in membrane binding (Davies 2014).

The N-terminal region varies between animal and plant. The N-terminal in an animal is longer in comparison to plants (plants have about 10 amino acid residues).

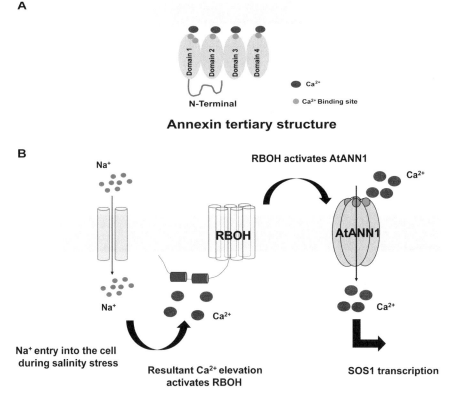

Fig. 8.1 The hypothetical tertiary structure of ANN and model describing the role of AtANN1 in Ca^{2+} signaling. (a) ANN with four repeat domains and N-terminal is depicted attached to the plasma membrane with its Ca^{2+} binding domain in the presence of Ca^{2+}. (This figure is adopted and modified according to Laohavisit and Davies 2009). **(b)** A hypothetical model describing the probable role played by AtANN1 during salinity stress. The Na^+ molecules upon its entry into the cytosol (during salt stress) cause elevation of cytosolic Ca^{2+} resulting in the activation of respiratory burst oxidase homolog (RBOH). As a result, ROS is produced which activates AtANN1 to mediate more Ca^{2+} release in the cytosol finally resulting in salt overly sensitive (SOS1) transcription. (This part of the figure is adopted and modified according to Davies 2014)

This region aids ANN and (other) protein interaction and supports different conformational states of ANN. This region can also undergo other modifications like phosphorylation, nitrosylation, S-glutathionylation, and N-myristoylation (Mortimer et al. 2008; Davies 2014). Nitrosylation and glutathionylation have also been reported for plant ANN (Lindermayr et al. 2005; Konopka-Postupolska et al. 2009). Similarly, phosphorylation has also been reported in plant ANN (Andrawis et al. 1993; Agrawal and Thelen 2006; Rohila et al. 2006; Wang et al. 2013). The S-glutathionylation decreases the Ca^{2+} affinity of ANN (Arabidopsis Annexin 1, literature uses two nomenclature AtANN1 or ANNAT1, we use the first one) and hence membrane association of ANN. For nitrosylation and phosphorylation, the functional aspect is still not clear.

Next, we come to the Annexin repeat domains (Domain I, Domain 2, Domain 3 and Domain 4). Each domain has roughly 70 amino acids and comprises of 5 α-helices (named A, B, C, D and E) joined by short loops (Laohavisit and Davies 2009). The loops are important for Ca^{2+} and ANN interaction (AB loop (connecting helix A to B) and DE loop (connecting helix D to E)) (Laohavisit and Davies 2011). Helix A, B, D and E are arranged in parallel and form a helix-loop-helix bundle and helix C is perpendicular to them (Konopka-Postupolska and Clark 2017). The Annexin domains perform two major functions- Ca^{2+} binding and membrane phospholipid binding. The Ca^{2+} binding task is performed by a type II Ca^{2+} binding motif (residues GxGT-(38–40 residues)-D/E, also known as an endonexin repeat) (Laohavisit and Davies 2009; Konopka-Postupolska and Clark 2017). This motif is present in two different α-helices and plants usually have this conserved motif in Domain 1. Domain 4 shows moderate conservation of this domain and Domain 2 and 3 are considered to have lost this motif (animals may have up to 3 or 4 endonexin motifs) (Laohavisit and Davies 2009; Konopka-Postupolska and Clark 2017). Even in the absence of typical endonexin domain, reports are suggesting that plant ANN can bind four Ca^{2+} ions (Hu et al. 2008). The ANN binds to negatively charged membrane phospholipids (phosphatidylserine (PS), phosphatidylinositol, and phosphatidic acid) in the presence of Ca^{2+} (Mortimer et al. 2008). The mammalian phospholipid (PS) binding motif is [R/K]XXXXDXXS[D/E]) (Montaville et al. 2002). This sequence is usually found in Domain 1 and sometimes in Domain 2 and overlaps the Ca^{2+}-binding endonexin domain (absent in Domains 3 and 4) (Mortimer et al. 2008). In plant ANNs, the sequence is not conserved although strict sequence conservation is not required for PS and membrane binding (Mortimer et al. 2008; Laohavisit and Davies 2011). Usually, the ANNs bind to the membrane depending on Ca^{2+}, however, other factors are influencing this event (like pH). The event can occur even in the absence of Ca^{2+} (Mortimer et al. 2008).

We now address one of the major functions of ANNs (besides membrane fusion), i.e., Ca^{2+} transport. Vertebrate ANNs (Annexin A7) can function as a voltage-gated channel (Pollard and Rojas 1988). Besides voltage; ATP, GTP, cAMP, hydrogen peroxide and low pH can all regulate ANNs (Mortimer et al. 2008). There are different views regarding channel formation by ANNs (reviewed in (Mortimer et al. 2008)). The "salt bridges" in the ANNs provide ion selectivity. The plant ANNs are known to transport Ca^{2+}, K^+, Na^+ and H^+ (Hofmann et al. 2000; Demidchik and

Maathuis 2007; Gorecka et al. 2007; Laohavisit et al. 2009, 2012, 2013). The mutant (and complementation) analysis of *Atann1* also supports Ca^{2+} conductance (Laohavisit et al. 2012; Richards et al. 2014). The AtANN1 functions in response to root salinity stress (hypothetical explanation in Fig. 8.1) (Laohavisit et al. 2013).

Besides these important domains, ANNs also have motifs for peroxidase activity and phosphodiesterase (PDE) activity. Many plant ANNs have shown peroxidase activity (Laohavisit and Davies 2009). The animal ANNs in comparison lack peroxidase activity (Laohavisit and Davies 2009). The S3 cluster is responsible for peroxidase activity (Laohavisit and Davies 2009, 2011). The peroxidase activity of ANNs might play certain roles such as peroxidation, helping in membrane localization of ANNs, and ANNs protecting membranes from peroxidation (Laohavisit and Davies 2009). The PDE activity, like peroxidase activity, is also a very common trait for plant ANNs (McClung et al. 1994; Calvert et al. 1996; Lim et al. 1998; Shin and Brown 1999). The putative GTP binding motif (GXXXXGKT and DXXG) is present in the Domain 4 (Clark et al. 2001). The function of ANN PDE activity is still speculative (Laohavisit and Davies 2009). ANNs are also able to bind to actin filaments, and only some plant ANNs show this property (Mortimer et al. 2008). A speculated IRI motif present in Domain 3 is responsible for actin binding (Laohavisit and Davies 2011), but plant ANNs with this domain may not show actin-binding (Mortimer et al. 2008). Lastly, some plant ANNs possess K-R-H-G-D motif (or some modifications of it) that allows interaction with C2- domain-containing proteins. These proteins may also help in Ca^{2+} dependent membrane binding (Laohavisit and Davies 2011).

ANNs are present in vertebrates, invertebrates, fungi, plants and protists (Moss and Morgan 2004; Mortimer et al. 2008; Jami et al. 2012). The plant ANNs generally show a trend of increase in number as we move from green algae to vascular plants (Clark et al. 2012) (In Fig. 8.2). The ANNs are found in growing cells (in root hairs and pollen) (Laohavisit and Davies 2009). The ANNs are ubiquitous in their presence in the plant cell- could be present in the PM, tonoplast, organellar membranes and also nucleolus (Laohavisit and Davies 2009). We have already mentioned the chief function of ANNs in membrane trafficking (reviewed extensively in (Konopka-Postupolska and Clark 2017)) and have discussed plant ANNs role in Ca^{2+} uptake. This is related to ANNs ability in modulating plants response to various stress and developmental responses (Mortimer et al. 2008; Demidchik and Shabala 2018; Yadav et al. 2018).

Mechanosensitive Channel

The MSc (also referred to as stretch-activated or force-gated channels) represent a special class of ion channels that respond to stimuli like touch, gravity, and osmotic pressure (different types of mechanical forces) (Hamilton et al. 2015; Basu and Haswell 2017). Even internal stimuli like (cell development, interactions between two cells) can also generate mechanical forces (and stimuli) (Toyota et al. 2018).

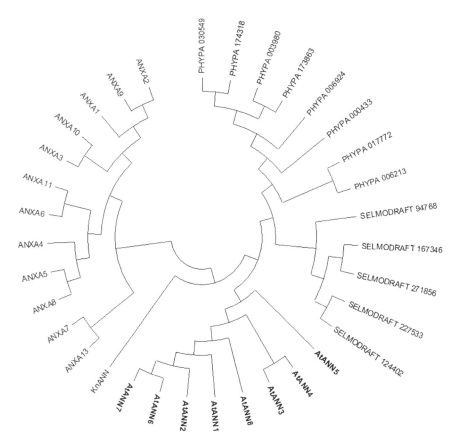

Fig. 8.2 Phylogenetic relationship of ANNs. A Neighbor-Joining tree is presented to show the rooted link to the *Homo sapiens* ANNs (marked in red). Each class of ANNs form separate groups. The *Arabidopsis thaliana* ANNs are marked in bold. The *Physcomitrella patens* (PHYPA), *Selaginella moellendorfii* (SELMODRAFT) and *Klebsormidium nitens* (Kn) ANNs were identified from UniProt database. Tree generated using MEGA10

The MSc have been identified in plants using different approaches (electrophysiology, channel blockers and mutant studies) (Hamilton et al. 2015). These channels are permeable to (Cl^-, Ca^{2+}, K^+) (Falke et al. 1988; Alexandre and Lassalles 1991; Ding and Pickard 1993; Spalding and Goldsmith 1993; Qi et al. 2004). There are also other reports of MSc channel activities for which permeability is unknown (reviewed in (Hamilton et al. 2015)). Besides, touch stimuli can result in Ca^{2+} rise in the cytosol, indicating that MSc are involved in Ca^{2+} transport (Knight et al. 1991). Further, two hyperosmolarity-induced $[Ca^{2+}]_{cyt}$ increase' channels (OSCA) (a class of MSc, discussed little later) are also involved in Ca^{2+} transport (Hou et al. 2014; Yuan et al. 2014). Summarizing all these pieces of evidence, the MSc channels are proposed to have permeability to Ca^{2+} (besides anion and K^+) (Demidchik et al. 2018).

Haswell and colleagues have forwarded some basic criteria for classifying ion channels as MSc- (a) their expression and localization, (b) their response to mechanical stimuli, but no significant role in normal plant development, (c) evidence of mechanosensitive gating (in the heterologous system) and (d) alterations in the protein structure perturbing the channel behavior (Hamilton et al. 2015). Based on these criteria (and allowing some freedom) the plants have four classes of MSc- Mechanosensitive-like channels (MSL), Mid1-complementing channels (MCA), OSCA (mentioned earlier) and two-pore potassium (TPK) families. There is another group known in plants- the Piezo-but at this moment their channel properties (and permeability) is unknown (Hamilton et al. 2015; Demidchik et al. 2018; Toyota et al. 2018). Currently, there are two models describing the gating (conformational change in response to stimuli, discussed for other channels in earlier chapters)- (1) lipid-disordering model, and (2) the hydrophobic mismatch (both models try to predict the gating in terms of the MSc and lipid bilayer interaction) (Hamilton et al. 2015).

The MSL channels are similar to the   *Escherichia coli* mechanosensitive channel of small conductance (MscS) (Wilson et al. 2013; Hamilton et al. 2015). The MSL channels in Arabidopsis are 10 in number and are divided into three different phylogenetic groups (Group 1, Group 2 and Group 3) (Hamilton et al. 2015). Group 1 and Group 2 MSL proteins have 5 transmembrane (TM) helices, the fifth helix in both is the pore-lining domain. The C-terminus is located in the mitochondrial matrix (for Group 1 MSL, its subcellular location) and chloroplast stroma (for Group 2 MSL, its subcellular location). Group 3 are different from the other members in having 6 TM helices (sixth becoming pore-lining domain), a large cytosolic N-terminus, a large loop connecting fourth and fifth helices and predominantly PM localization (Hamilton et al. 2015). They exist as heptamers or pentamers (Demidchik et al. 2018). There are several pieces of evidence in favor of ion channel activity of MSL channels (Haswell and Meyerowitz 2006; Haswell et al. 2008; Peyronnet et al. 2008; Maksaev and Haswell 2012). MSLs can maintain a potential difference across mitochondria, ion balance, mechanical sensing and programmed cell death (Peyronnet et al. 2008; Veley et al. 2014; Lee et al. 2016; Hamilton and Haswell 2017; Basu and Haswell 2020). Besides these, there are other important roles of MSLs in plants (reviewed in (Hamilton et al. 2015; Demidchik et al. 2018)).

The MCAs are thought to be important MSc contributing to cellular Ca^{2+} influx (Hamilton et al. 2015). The plant MCA were identified by complementation of yeast Mid1 protein (an MSc involved in Ca^{2+} influx) mutant and hence the nomenclature "Mid1-complementing" (Hamilton et al. 2015). The MCA are single TM helix containing proteins that form homo tetramers (Shigematsu et al. 2014). The MCA has an EF-hand like motif at N-terminus, followed by the single TM helix, and then the C-terminus (has a coiled-coil motif and Plac8 motif) (Hamilton et al. 2015; Kamano et al. 2015). The N-terminal half with the EF-hand-like region modulates Ca^{2+}

uptake and that the coiled-coil motif differently regulates two MCAs of Arabidopsis (data inferred from heterologous experiments) (Nakano et al. 2011). Like the MSL channels, electrophysiological analysis has proved MCAs are also ion channels (Furuichi et al. 2012). The MCAs are localized to the PM (Hamilton et al. 2015). In contrast to MSL whose Ca^{2+} conductance is still debatable (Demidchik et al. 2018), there are several pieces of evidence for the involvement of MCAs in Ca^{2+} homeostasis (Nakagawa et al. 2007; Yamanaka et al. 2010; Kurusu et al. 2012; Iida et al. 2014).

The OSCA respond to osmotic stress in response to salinity and stress and trigger Ca^{2+} influx into the cytosol (Hou et al. 2014; Yuan et al. 2014; Murthy et al. 2018). OSCA structures have been determined recently for *Arabidopsis thaliana* OSCA1.2 and OSCA1.3 and *Oryza sativa* OSCA1.2 (Jojoa-Cruz et al. 2018; Liu et al. 2018; Zhang et al. 2018; Maity et al. 2019). There were predicted 11 TM helices in OSCA and it forms a homodimer. The loop, present in the cytosolic portion, connecting TM3 and TM4 is large and sometimes mentioned as CTD (the numbering of the helices are changed to TM2 and TM3 in literature, depending on the numbering of the first helix, which we have named TM1) (Zhang et al. 2018). The pore helices are formed by five TM helices (TM4 to TM8). The conformational change involves TM1 to TM7. Several residues are involved in the ion conductance (Jojoa-Cruz et al. 2018; Liu et al. 2018; Zhang et al. 2018; Maity et al. 2019). The OSCA are reported to be localized to the PM and 15 members are present in *Arabidopsis thaliana* (Murthy et al. 2018). Figure 8.3 shows the predictive membrane topology of major plant MSc.

The TPKs are K^+ selective MSc and can be modulated by the cytosolic Ca^{2+}. They are mostly localized to the vacuole (Hamilton et al. 2015). The plant Piezo channels are far less characterized in the MSc group (compared to animals). At this moment, there is no experimental data for plant Piezo channel activity or ion selectivity (Demidchik et al. 2018). Figure 8.4 shows the phylogenetic relationship of major plant MSc involved (or putatively involved) in Ca^{2+} influx.

Conclusion

The ANNs and MSc are not typical ion channels but serve the purpose of Ca^{2+} influx into the plant cell. Of special mention is the MSc that are currently being characterized and have opened a new doorway to understand plant Ca^{2+} influx.

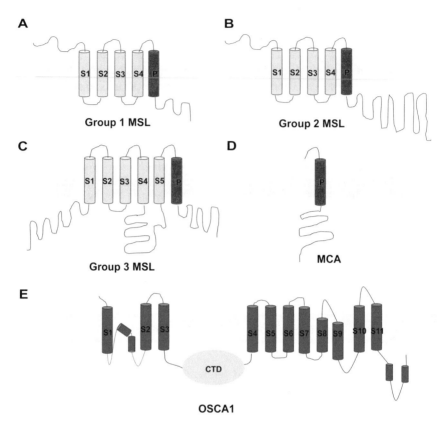

Fig. 8.3 Hypothetical structure of mechanosensitive channels. Mechanosensitive-like channels (MSL) in Arabidopsis have three groups (group 1, group 2 and group 3-as **a**, **b** and **c**, respectively). The helix marked in red is the pore region (according to Demidchik et al. 2018). The N-terminal and C-terminal also differ (according to Hamilton et al. 2015). Mid1-complementing channels (MCA as **d**) channel has only a single transmembrane helix, which is also the pore region. The reduced 'hyperosmolarity-induced [Ca^{2+}]$_{cyt}$ increase' channels (OSCA1 as **e**) with 11 transmembrane helices. A large cytosolic CTD between S3 and S4 helix. The MSL figures are adapted from both (Hamilton et al. 2015; Demidchik et al. 2018). The MCA channel is adopted from Demidchik et al. (2018) and OSCA from Zhang et al. (2018)

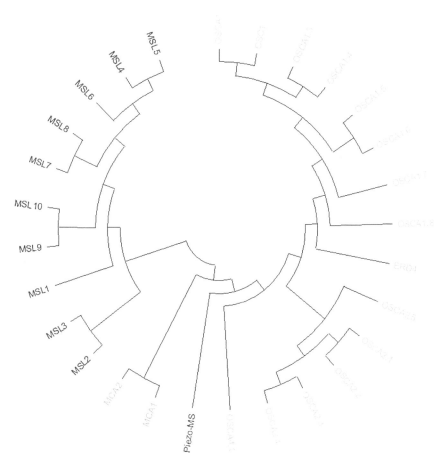

Fig. 8.4 Phylogenetic relationship of *Arabidopsis thaliana* Mechanosensitive channel. An unrooted Neighbor-Joining tree is presented of MSL (10 in total, marked in red), MCA (2 in total, marked in red), Piezo (1 in total, marked in violet) and OSCA (15 in total, marked in green). Sequences were identified from UniProt database. Tree generated using MEGA10

References

G.K. Agrawal, J.J. Thelen, Large scale identification and quantitative profiling of Phosphoproteins expressed during seed filling in oilseed rape. Mol. Cell. Proteomics **5**, 2044–2059 (2006)

J. Alexandre, J.P. Lassalles, Hydrostatic and osmotic pressure activated channel in plant vacuole. Biophys. J. **60**, 1326–1336 (1991)

A. Andrawis, M. Solomon, D.P. Delmer, Cotton fiber annexins: A potential role in the regulation of callose synthase. Plant J. **3**, 763–772 (1993)

D. Basu, E.S. Haswell, Plant Mechanosensitive ion channels: An ocean of possibilities. Curr. Opin. Plant Biol. **40**, 43–48 (2017)

D. Basu, E.S. Haswell, The Mechanosensitive Ion Channel MSL10 potentiates responses to cell swelling in Arabidopsis seedlings. Curr. Biol **30**, 2716–2728 (2020)

C.M. Calvert, S.J. Gant, D.J. Bowles, Tomato annexins p34 and p35 bind to F-actin and display nucleotide phosphodiesterase activity inhibited by phospholipid binding. Plant Cell **8**, 333–342 (1996)

G.B. Clark, A. Sessions, D.J. Eastburn, S.J. Roux, Differential expression of members of the Annexin multigene family in Arabidopsis. Plant Physiol. **126**, 1072–1084 (2001)

G.B. Clark, R.O. Morgan, M.P. Fernandez, S.J. Roux, Evolutionary adaptation of plant annexins has diversified their molecular structures, interactions and functional roles. New Phytol. **196**, 695–712 (2012)

J.M. Davies, Annexin-mediated calcium Signalling in plants. Plants (Basel) **3**, 128–140 (2014)

V. Demidchik, F.J. Maathuis, Physiological roles of nonselective cation channels in plants: From salt stress to signalling and development. New Phytol. **175**, 387–404 (2007)

V. Demidchik, S. Shabala, Mechanisms of cytosolic calcium elevation in plants: The role of ion channels, calcium extrusion systems and NADPH oxidase-mediated 'ROS-Ca(2+) Hub'. Funct. Plant Biol. **45**, 9–27 (2018)

V. Demidchik, S. Shabala, S. Isayenkov, T.A. Cuin, I. Pottosin, Calcium transport across plant membranes: Mechanisms and functions. New Phytol. **220**, 49–69 (2018)

J.P. Ding, B.G. Pickard, Mechanosensory calcium-selective cation channels in epidermal cells. Plant J. **3**, 83–110 (1993)

L.C. Falke, K.L. Edwards, B.G. Pickard, S. Misler, A stretch-activated anion channel in tobacco protoplasts. FEBS Lett. **237**, 141–144 (1988)

T. Furuichi, H. Iida, M. Sokabe, H. Tatsumi, Expression of Arabidopsis MCA1 enhanced mechanosensitive channel activity in the Xenopus laevis oocyte plasma membrane. Plant Signal. Behav. **7**, 1022–1026 (2012)

K.M. Gorecka, C. Thouverey, R. Buchet, S. Pikula, Potential role of annexin AnnAt1 from Arabidopsis thaliana in pH-mediated cellular response to environmental stimuli. Plant Cell Physiol. **48**, 792–803 (2007)

E.S. Hamilton, E.S. Haswell, The tension-sensitive ion transport activity of MSL8 is critical for its function in pollen hydration and germination. Plant Cell Physiol. **58**, 1222–1237 (2017)

E.S. Hamilton, A.M. Schlegel, E.S. Haswell, United in diversity: Mechanosensitive ion channels in plants. Annu. Rev. Plant Biol. **66**, 113–137 (2015)

E.S. Haswell, E.M. Meyerowitz, MscS-like proteins control plastid size and shape in Arabidopsis thaliana. Curr. Biol. **16**, 1–11 (2006)

E.S. Haswell, R. Peyronnet, H. Barbier-Brygoo, E.M. Meyerowitz, J.M. Frachisse, Two MscS homologs provide mechanosensitive channel activities in the Arabidopsis root. Curr. Biol. **18**, 730–734 (2008)

A. Hofmann, J. Proust, A. Dorowski, R. Schantz, R. Huber, Annexin 24 from Capsicum annuum. X-ray structure and biochemical characterization. J. Biol. Chem. **275**, 8072–8082 (2000)

A. Hofmann, S. Ruvinov, S. Hess, R. Schantz, D.P. Delmer, A. Wlodawer, Plant annexins form calcium-independent oligomers in solution. Protein Sci. **11**, 2033–2040 (2002)

C. Hou, W. Tian, T. Kleist, K. He, V. Garcia, F. Bai, Y. Hao, S. Luan, L. Li, DUF221 proteins are a family of osmosensitive calcium-permeable cation channels conserved across eukaryotes. Cell Res. **24**, 632–635 (2014)

N.J. Hu, A.M. Yusof, A. Winter, A. Osman, A.K. Reeve, A. Hofmann, The crystal structure of calcium-bound annexin Gh1 from Gossypium hirsutum and its implications for membrane binding mechanisms of plant annexins. J. Biol. Chem. **283**, 18314–18322 (2008)

H. Iida, T. Furuichi, M. Nakano, M. Toyota, M. Sokabe, H. Tatsumi, New candidates for mechano-sensitive channels potentially involved in gravity sensing in Arabidopsis thaliana. Plant Biol. (Stuttg.) **16**(Suppl 1), 39–42 (2014)

S.K. Jami, G.B. Clark, B.T. Ayele, P. Ashe, P.B. Kirti, Genome-wide comparative analysis of Annexin superfamily in plants. PLoS One **7**, e47801 (2012)

S. Jojoa-Cruz, K. Saotome, S.E. Murthy, C.C.A. Tsui, M.S. Sansom, A. Patapoutian, A.B. Ward, Cryo-EM structure of the mechanically activated ion channel OSCA1.2. elife **7**, e41845 (2018)

S. Kamano, S. Kume, K. Iida, K.J. Lei, M. Nakano, Y. Nakayama, H. Iida, Transmembrane topologies of Ca2+−permeable Mechanosensitive channels MCA1 and MCA2 in Arabidopsis thaliana. J. Biol. Chem. **290**, 30901–30909 (2015)

M.R. Knight, A.K. Campbell, S.M. Smith, A.J. Trewavas, Transgenic plant aequorin reports the effects of touch and cold-shock and elicitors on cytoplasmic calcium. Nature **352**, 524–526 (1991)

D. Konopka-Postupolska, G. Clark, Annexins as overlooked regulators of membrane trafficking in plant cells. Int. J. Mol. Sci. **18**, 863 (2017)

D. Konopka-Postupolska, G. Clark, G. Goch, J. Debski, K. Floras, A. Cantero, B. Fijolek, S. Roux, J. Hennig, The role of Annexin 1 in drought stress in Arabidopsis1. Plant Physiol. **150**, 1394–1410 (2009)

T. Kurusu, D. Nishikawa, Y. Yamazaki, M. Gotoh, M. Nakano, H. Hamada, T. Yamanaka, K. Iida, Y. Nakagawa, H. Saji, K. Shinozaki, H. Iida, K. Kuchitsu, Plasma membrane protein OsMCA1 is involved in regulation of hypo-osmotic shock-induced Ca2+ influx and modulates generation of reactive oxygen species in cultured rice cells. BMC Plant Biol. **12**, 11 (2012)

A. Laohavisit, J.M. Davies, Multifunctional annexins. Plant Sci. **177**, 532–539 (2009)

A. Laohavisit, J.M. Davies, Annexins. New Phytol. **189**, 40–53 (2011)

A. Laohavisit, J.C. Mortimer, V. Demidchik, K.M. Coxon, M.A. Stancombe, N. Macpherson, C. Brownlee, A. Hofmann, A.A. Webb, H. Miedema, N.H. Battey, J.M. Davies, Zea mays Annexins modulate cytosolic free Ca2+ and generate a Ca2+-permeable conductance. Plant Cell **21**, 479–493 (2009)

A. Laohavisit, Z. Shang, L. Rubio, T.A. Cuin, A.A. Véry, A. Wang, J.C. Mortimer, N. Macpherson, K.M. Coxon, N.H. Battey, C. Brownlee, O.K. Park, H. Sentenac, S. Shabala, A.A. Webb, J.M. Davies, Arabidopsis Annexin1 mediates the radical-activated plasma membrane Ca2+− and K+-permeable conductance in root cells. Plant Cell **24**, 1522–1533 (2012)

A. Laohavisit, S.L. Richards, L. Shabala, C. Chen, R.D. Colaço, S.M. Swarbreck, E. Shaw, A. Dark, S. Shabala, Z. Shang, J.M. Davies, Salinity-induced calcium signaling and root adaptation in Arabidopsis require the calcium regulatory protein Annexin11. Plant Physiol. **163**, 253–262 (2013)

C.P. Lee, G. Maksaev, G.S. Jensen, M.W. Murcha, M.E. Wilson, M. Fricker, R. Hell, E.S. Haswell, A.H. Millar, L. Sweetlove, MSL1 is a mechanosensitive ion channel that dissipates mitochondrial membrane potential and maintains redox homeostasis in mitochondria during abiotic stress. Plant J. **88**, 809–825 (2016)

E.K. Lim, M.R. Roberts, D.J. Bowles, Biochemical characterization of tomato annexin p35. Independence of calcium binding and phosphatase activities. J. Biol. Chem. **273**, 34920–34925 (1998)

C. Lindermayr, G. Saalbach, J. Durner, Proteomic identification of S-Nitrosylated proteins in Arabidopsis. Plant Physiol. **137**, 921–930 (2005)

X. Liu, J. Wang, L. Sun, Structure of the hyperosmolality-gated calcium-permeable channel OSCA1.2. Nat. Commun. **9**, 5060 (2018)

K. Maity, J.M. Heumann, A.P. McGrath, N.J. Kopcho, P.K. Hsu, C.W. Lee, J.H. Mapes, D. Garza, S. Krishnan, G.P. Morgan, K.J. Hendargo, T. Klose, S.D. Rees, A. Medrano-Soto, M.H. Saier, M. Piñeros, E.A. Komives, J.I. Schroeder, G. Chang, M.H.B. Stowell, Cryo-EM structure of OSCA1.2 from Oryza sativa elucidates the mechanical basis of potential membrane hyperosmolality gating. Proc. Natl. Acad. Sci. U. S. A. **116**, 14309–14318 (2019)

G. Maksaev, E.S. Haswell, MscS-Like10 is a stretch-activated ion channel from Arabidopsis thaliana with a preference for anions. Proc. Natl. Acad. Sci. U. S. A. **109**, 19015–19020 (2012)

A.D. McClung, A.D. Carroll, N.H. Battey, Identification and characterization of ATPase activity associated with maize (Zea mays) annexins. Biochem. J. **303**, 709–712 (1994)

P. Montaville, J.M. Neumann, F. Russo-Marie, F. Ochsenbein, A. Sanson, A new consensus sequence for phosphatidylserine recognition by annexins. J. Biol. Chem. **277**, 24684–24693 (2002)

J.C. Mortimer, A. Laohavisit, N. Macpherson, A. Webb, C. Brownlee, N.H. Battey, J.M. Davies, Annexins: Multifunctional components of growth and adaptation. J. Exp. Bot. **59**, 533–544 (2008)

S.E. Moss, R.O. Morgan, The annexins. Genome Biol. **5**, 219 (2004)

S.E. Murthy, A.E. Dubin, T. Whitwam, S. Jojoa-Cruz, S.M. Cahalan, S.A.R. Mousavi, A.B. Ward, A. Patapoutian, OSCA/TMEM63 are an evolutionarily conserved family of mechanically activated ion channels. elife **7**, e41844 (2018)

Y. Nakagawa, T. Katagiri, K. Shinozaki, Z. Qi, H. Tatsumi, T. Furuichi, A. Kishigami, M. Sokabe, I. Kojima, S. Sato, T. Kato, S. Tabata, K. Iida, A. Terashima, M. Nakano, M. Ikeda, T. Yamanaka, H. Iida, Arabidopsis plasma membrane protein crucial for Ca2+ influx and touch sensing in roots. Proc. Natl. Acad. Sci. U. S. A. **104**, 3639–3644 (2007)

M. Nakano, K. Iida, H. Nyunoya, H. Iida, Determination of structural regions important for Ca(2+) uptake activity in Arabidopsis MCA1 and MCA2 expressed in yeast. Plant Cell Physiol. **52**, 1915–1930 (2011)

R. Peyronnet, E.S. Haswell, H. Barbier-Brygoo, J.M. Frachisse, AtMSL9 and AtMSL10: Sensors of plasma membrane tension in Arabidopsis roots. Plant Signal. Behav. **3**, 726–729 (2008)

H.B. Pollard, E. Rojas, Ca2+−activated synexin forms highly selective, voltage-gated Ca2+ channels in phosphatidylserine bilayer membranes. Proc. Natl. Acad. Sci. U. S. A. **85**, 2974–2978 (1988)

Z. Qi, A. Kishigami, Y. Nakagawa, H. Iida, M. Sokabe, A mechanosensitive anion channel in Arabidopsis thaliana mesophyll cells. Plant Cell Physiol. **45**, 1704–1708 (2004)

S.L. Richards, A. Laohavisit, J.C. Mortimer, L. Shabala, S.M. Swarbreck, S. Shabala, J.M. Davies, Annexin 1 regulates the H2O2-induced calcium signature in Arabidopsis thaliana roots. Plant J. **77**, 136–145 (2014)

J.S. Rohila, M. Chen, S. Chen, J. Chen, R. Cerny, C. Dardick, P. Canlas, X. Xu, M. Gribskov, S. Kanrar, J.K. Zhu, P. Ronald, M.E. Fromm, Protein-protein interactions of tandem affinity purification-tagged protein kinases in rice. Plant J. **46**, 1–13 (2006)

H. Shigematsu, K. Iida, M. Nakano, P. Chaudhuri, H. Iida, K. Nagayama, Structural characterization of the Mechanosensitive Channel candidate MCA2 from Arabidopsis thaliana. PLoS One **9**, e87724 (2014)

H. Shin, R.M. Brown, GTPase activity and biochemical characterization of a recombinant cotton Fiber Annexin. Plant Physiol. **119**, 925–934 (1999)

E. Spalding, M. Goldsmith, Activation of K+ channels in the plasma membrane of Arabidopsis by ATP produced Photosynthetically. Plant Cell **5**, 477–484 (1993)

M. Toyota, T. Furuichi, H. Iida, Molecular Mechanisms of Mechanosensing and Mechanotransduction, in *Plant Biomechanics: From Structure to Function at Multiple Scales*, ed. by A. Geitmann, J. Gril, (Springer, Cham, 2018), pp. 375–397

K.M. Veley, G. Maksaev, E.M. Frick, E. January, S.C. Kloepper, E.S. Haswell, Arabidopsis MSL10 has a regulated cell death signaling activity that is separable from its Mechanosensitive Ion Channel activity. Plant Cell **26**, 3115–3131 (2014)

P. Wang, L. Xue, G. Batelli, S. Lee, Y.J. Hou, M.J. Van Oosten, H. Zhang, W.A. Tao, J.K. Zhu, Quantitative phosphoproteomics identifies SnRK2 protein kinase substrates and reveals the effectors of abscisic acid action. Proc. Natl. Acad. Sci. U. S. A. **110**, 11205–11210 (2013)

M.E. Wilson, G. Maksaev, E.S. Haswell, MscS-like Mechanosensitive channels in plants and microbes. Biochemistry **52**, 5708–5722 (2013)

D. Yadav, P. Boyidi, I. Ahmed, P.B. Kirti, Plant annexins and their involvement in stress responses. Environ. Exp. Bot. **155**, 293–306 (2018)

T. Yamanaka, Y. Nakagawa, K. Mori, M. Nakano, T. Imamura, H. Kataoka, A. Terashima, K. Iida, I. Kojima, T. Katagiri, K. Shinozaki, H. Iida, MCA1 and MCA2 that mediate Ca2+ uptake have distinct and overlapping roles in Arabidopsis. Plant Physiol. **152**, 1284–1296 (2010)

F. Yuan, H. Yang, Y. Xue, D. Kong, R. Ye, C. Li, J. Zhang, L. Theprungsirikul, T. Shrift, B. Krichilsky, D.M. Johnson, G.B. Swift, Y. He, J.N. Siedow, Z.M. Pei, OSCA1 mediates osmotic-stress-evoked Ca2+ increases vital for osmosensing in Arabidopsis. Nature **514**, 367–371 (2014)

M. Zhang, D. Wang, Y. Kang, J.X. Wu, F. Yao, C. Pan, Z. Yan, C. Song, L. Chen, Structure of the mechanosensitive OSCA channels. Nat. Struct. Mol. Biol. **25**, 850–858 (2018)

Chapter 9
Ca²⁺-ATPase and Ca²⁺/Cation Antiporters

Contents

Introduction

In the last few chapters, we have discussed the "transport elements" that bring Ca^{2+} into the cytoplasm, increasing the cytosolic Ca^{2+} concentration by several folds. As we have mentioned in Chap. 3, there are very efficient mechanisms present in the plant cell to maintain the Ca^{2+} homeostasis, by removing the cytosolic Ca^{2+} and putting it back to the Ca^{2+} storing reservoirs. In this chapter, we discuss the P-Type ATPase pump (the Ca^{2+}-ATPase) and Ca^{2+}/cation antiporter (CaCA) superfamily (Palmgren and Nissen 2011; Pittman and Hirschi 2016b). At this juncture, we make it clear to the readers that –(a) each member of these superfamilies is not involved in Ca^{2+} transport and, (b) not all the members are present in plants.

P-Type ATPases

The P-Type ATPases use ATP hydrolysis to mediate ion/counter ion transport (Palmgren and Nissen 2011). There are five different ATPases families (P1, P2, P3, P4 and P5) further subdivided into subclasses (A, B, C). Figure 9.1 shows a

G. K. Pandey, S. K. Sanyal, *Functional Dissection of Calcium Homeostasis and Transport Machinery in Plants*, SpringerBriefs in Plant Science, https://doi.org/10.1007/978-3-030-58502-0_9

phylogenetic tree depicting the different ATPases (11 in total- P1A, P1B, P2A, P2B, P2C, P2D, P3A, P3B, P4, P5A and P5B). The P1A is involved in high-affinity uptake of K⁺, P1B in pumping heavy metal, P2A is ER-Type Ca²⁺-ATPases (ECA) and P2B are autoinhibited Ca²⁺-ATPases (ACA)- both of these are involved majorly in Ca²⁺ transport, P2C is Na⁺/K⁺-ATPases (and H⁺/K⁺ pumps of animals), P2D are Na⁺ or K⁺ pumps, P3A are autoinhibited H⁺-ATPases, P3B is Mg²⁺-ATPases, P4 are phospholipid flippases and P5 (A and B) are ATPases with unknown function (Palmgren and Nissen 2011; Pedersen et al. 2012; Huda et al. 2013). The P1A and P3B are exclusively prokaryotic. *Arabidopsis thaliana* has P1B, P2 (A and B), P3A, P4 and P5A present in its genome. *Selaginella moelendorfii* has P5B (besides the ones mentioned for *A. thaliana*). *Physcomitrella patens* have P2D (besides the ones mentioned for the former two). *Ostreococcus tauri* has P2C (and others of plant lineage, except P2D) (Pedersen et al. 2012). The animal and plant genome shows some marked differences in the presence of ATPases- especially the P2C, P2D and P3A classes. Animals have P2C (also present in chlorophytes like *O. tauri* and *Chlamydomonas reinharditii*), but lack P2D (present in algae and moss *P. patens*) and P3A (present in all plant lineage) (Pedersen et al. 2012). Summarizing from this

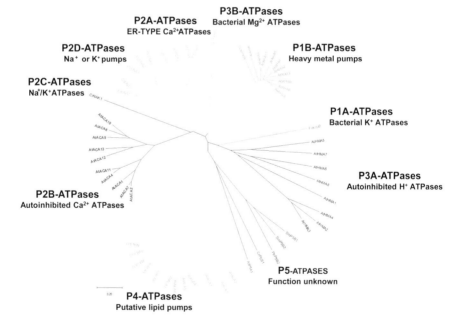

Fig. 9.1 A phylogenetic tree depicting the members of P-type ATPases. The unrooted Neighbor-Joining tree depicts family members present in *Arabidopsis thaliana* (At) (P1B, P2A, P2B, P3A, P4 and P5). The other members (P1A, P2C, P2D and P3A) sequences have been taken from *Escherichia coli* (Ec), *Chlamydomonas reinharditii* (Cr) and *Physcomitrela Patens* (Pp). P5 has both P5A and P5B (from *Selaginella moelendorfii* (Sm)). The sequences were taken from Pedersen et al. (2012) and UniProt and classification was done according to Palmgren and Nissen (2011), Pedersen et al. (2012), Huda et al. (2013). Tree constructed using MEGA10

information, the ATPases play a crucial role in modulating the energization of the cell (Na^+ in animal versus H^+ in the higher plant). Except for these differences (P2C, P2D and P3A) and the absentees (P1A and P3B), the other ATPases are common among animals and plants. Since the P2A and P2B (henceforth in the chapter ECA and ACA, respectively) are the only members involved in Ca^{2+} transport, our discussion from here on will be focused on these two classes of ATPases (unless specifically mentioned).

The Structure and Transport Mechanism of ATPases

The structure and transport mechanism of ATPases are remarkably similar even though there are significant differences in their sequences (Palmgren and Nissen 2011). An ATPase can be divided into the membrane-embedded domain (TM domain-TMD) and cytosolic region. The pump function due to coordination between these two domains. An ATP hydrolysis event is mediated by the certain domains in the cytoplasmic side and, ion binding and resultant movements in the TMD result into the transport of Ca^{2+} (and counter transport of H^+) (Palmgren and Nissen 2011). The cytoplasmic domain has an Actuator domain (AcD), Phosphorylation domain (PD) and Nucleotide-binding domain (ND) (Morth et al. 2011; Palmgren and Nissen 2011). AcD is the phosphatase domain with a signature motif TGE (Thr-Gly-Glu). The PD has the signature motif with DKTG (Asp-Lys-Thr-Gly) and the Asp (D) is phosphorylated to power the transport. The ND has a signature domain (signature motif KGAPE) and helps in ATP binding and phosphorylation of PD (Thever and Saier 2009; Huda et al. 2013). Besides these conserved domains in the cytoplasmic side, ACA has an N-terminal autoinhibitor domain (AD) (Tidow et al. 2012). It inhibits the pump and is activated in response to Calmodulin (CaM) binding. The TMD has 10 helices with the first 6 constituting the transport domain (TD) responsible, majorly, for the ion transport (for this reason the helices are flexible) (Palmgren and Nissen 2011). There are two ion-binding sites in ECA and one in ACA. ACA is selective and transports only Ca^{2+} and ECA can additionally mediate transport of Mn^{2+}, Cd^{2+} and Zn^{2+} (Demidchik et al. 2018). The next 4 TM helices constitute the support domain (SD) and provide structural support and auxiliary ion-conducting side chain. These helixs are rigid compared to TD helices and do not move during ion transport (Palmgren and Nissen 2011). Figure 9.2 briefly illustrates the membrane topology of ACA and ECA with the domains discussed above.

The ion transport mechanism of ATPase follows Albers model (Morth et al. 2011). A catalytic cycle involves hydrolysis of an ATP molecule resulting in the transport (and counter transport) of ions. Ion binding (Ca^{2+}) into the high-affinity site results into the phosphorylation of the PD (using an ATP molecule and action of ND) resulting in the E1P state of the pump. The pump oscillates from the phosphorylated to dephosphorylated state and involves the cytoplasmic domain (AcD, ND and PD) and the TD and their co-ordination finally removes Ca^{2+} from the

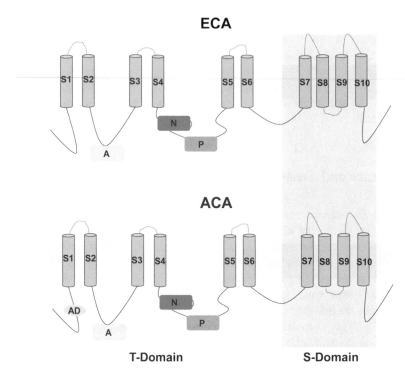

Fig. 9.2 Topological model of plant ACA and ECA. A hypothetical structure of the ACA and ECA pumps with its domains. The cytoplasmic common domains are Actuator domain (A), Phosphorylation domain (P) and Nucleotide-binding domain (N). The ACA has an extra autoinhibitory domain (AD). The 10 transmembrane domains are divided into transport domain (T-Domain) and support domain (S-Domain). (Adopted and modified from Demidchik et al. 2018)

cytosol (Morth et al. 2011; Palmgren and Nissen 2011). The different stages of this event are explained in Fig. 9.3.

Regulation of ACA and ECA

The AD domain of ACA, as the name suggests, keeps the ACA pump in an auto-inhibited state. The AD domain is regulated by Ca^{2+} and CaM (somewhat like the CNGC). There are two CaM binding sites in ACA, overlap the AD domain (CaM binding site (CaMBS1) and CaMBS2), one after the other (in ACA8, they are separated by 8 amino acids) (Tidow et al. 2012). In basal condition (when the cytosolic Ca^{2+} is at resting level), the AD sits over the catalytic core (AD, PD and ND) of ACA and keeps the pump auto-inhibited. Increasing Ca^{2+} in the cytoplasm mediates Ca^{2+} (bounded) CaM binding to CaMBS1. As Ca^{2+} concentration further increases, another CaM (bounded to Ca^{2+}) binds to CaMBS2. When both sites are loaded with CaM, the AD is moved away from the catalytic core of ACA, and the pump is now

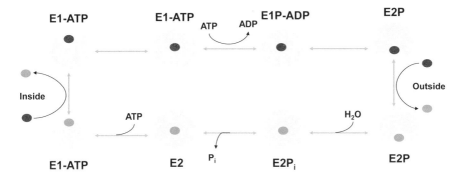

A Catalytic cycle of a P-type ATPase

● The ion that is to be transported

○ Counterion

Fig. 9.3 A hypothetical model explaining the catalytic cycle of P-type ATPases. The ATPase pump utilizes ATP hydrolysis to power ion transport across the membrane. The entire cycle is called a catalytic cycle and starts with binding of an ion to the E1-ATP state. The E1 state has a high affinity for Ca^{2+}. This leads to phosphorylation and resultant closed-form E1-ADP. The pump gradually moves to E2 state, which has less affinity for the ion (already bound) and more affinity for counterion, which it now binds. A subsequent dephosphorylation event forms a closed E2 state. Further ATP binding will release the counterion inside the cell. P2A exports 2 Ca^{2+} and counter transports H^+ and P2B exports one Ca^{2+} and counter transports one H^+ (Palmgren and Nissen 2011). (Adopted and modified from Morth et al. (2011)

active to pump out Ca^{2+} from the cytosol (Tidow et al. 2012). Figure 9.4 gives a brief description of the CaMBS sites in the AD and a model describing the modulation of ACA pumps by CaM. Besides this mode, preliminary data suggests that ACA can be phosphoregulated by kinases (ACA8 by CIPK9, Ca^{2+}-dependent protein kinase (CDPK aka CPK) (CPK1 and CPK16) and ACA2 by CDPK (CPK1)), but data at this point is not sufficient to predict a mechanistic model (Hwang et al. 2000; Giacometti et al. 2012; Costa et al. 2017). The phosphorylation by CDPK inhibits basal activity of ACA2 (Hwang et al. 2000). The more detailed analysis of ACA8 (using mutant ACA8 mimicking phosphorylated and dephosphorylated Ser residues) that phosphorylation affected the autoinhibition action of the N-terminus. It also changed the activation and de-activation kinetics modulated by CaM on ACA8 (by affecting its affinity for CaM) (Giacometti et al. 2012). It has been proven *in vivo* (using *in vivo* Ca^{2+} measurement reporter system) that CIPK9 mediated phosphorylation of ACA8 changes the Ca^{2+} signature dynamics (Costa et al. 2017), thus verifying the role of phosphorylation in modulating Ca^{2+} extrusion by ACA. The regulation of ECA by the above-mentioned model is not proven (further ECA lacks a regulatory domain like AD of ACA).

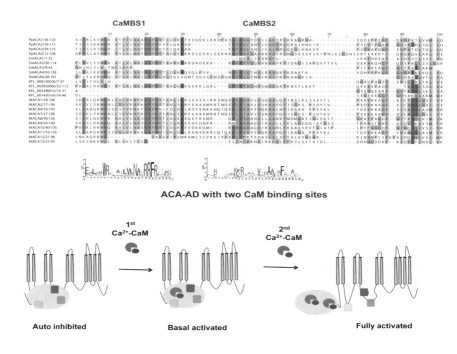

Fig. 9.4 A hypothetical model explaining the modulation of ACA ATPases by Ca²⁺-CaM. The auto-inhibitory domain (AD) with the two CaMBS1 are represented. Sequences from *Arabidopsis thaliana* (At), *Physcomitrela Patens* (Pp), *Selaginella moelendorfii* (Sm), and *Klebsormidium nitens* (KFL) were used for analysis. First CaMBS1 has the typical "RRFR" motif (in consensus motif). Alignment performed using MEGA, visualized using Jalview and alignment logo generated using WebLogo. The motif was not present in all the analyzed sequences. Sequences for AD and CaMBS are from Tidow et al. (2012). The effect of increasing Ca²⁺ concentration on modulating the ACA pump inhibition. Ca²⁺-CaM binding can remove autoinhibition. (Adapted and modified from Tidow et al. 2012)

Evolution, Localization and Biological Function of ACA and ECA

The ECA and ACA are found in most of the plant species and usually, ACAs dominate the ECA numerically (in Arabidopsis 10 ACA and 4 ECA) (Huda et al. 2013; Singh et al. 2014). The ECA is closely related to the animal sarco/endoplasmic reticulum Ca²⁺-ATPase (SERCA) pumps (Altshuler et al. 2012). It has been shown that gene duplication events increased the number of SERCA pumps. So the plants had two different clades present in their genome (Arabidopsis ECA1, ECA2 and ECA4, clade 1 and ECA3 in clade 2) (Altshuler et al. 2012). The specialized ECA functioning in fungi and animal Golgi, secretory pathway calcium pump (SPCA) is present only in *P. patens* and absent in other plants (Pedersen et al. 2012). Further, we have already mentioned the difference in the positioning of AD in ACA in animals and plants. Summarizing, we can say that the ECA and ACA families evolved to serve the specific needs of the plant and animals. Figure 9.5 summarizes the phylogenetic relationship in plant ECA and ACA showing evolutionary relatedness among the proteins belonging to Arabidopsis, *Physcomitrela*, *Selaginella*, *Klebsormidium*, and *Oryza*. The evolutionary relatedness for Ca^{2+} ATPases between the monocot (*Oryza)* and dicot (Arabidopsis) has been previously described (Singh et al. 2014). The ECA is majorly reported in the endoplasmic reticulum (ER) and Golgi (Wu et al. 2002; Mills et al. 2008; Huda et al. 2013). ECA from tomato was reported in both plasma membrane (PM) and tonoplast (Wimmers et al. 1992). Some ECA type activity (molecular identity unknown) was also reported at the PM (Thomson et al. 1993). Compared to the ECA, ACA is more distributed in the cell. They have been detected in the chloroplast, ER, small vacuole, PM (Huang et al. 1993; Hong et al. 1999; Axelsen and Palmgren 2001; Schiott et al. 2004; Boursiac et al. 2010; Frei dit Frey et al. 2012; Lucca and León 2012).

Both the ACA and ECA have been implicated in plant stress responses (abiotic and biotic) (reviewed in (Huda et al. 2013; Demidchik et al. 2018). They have also

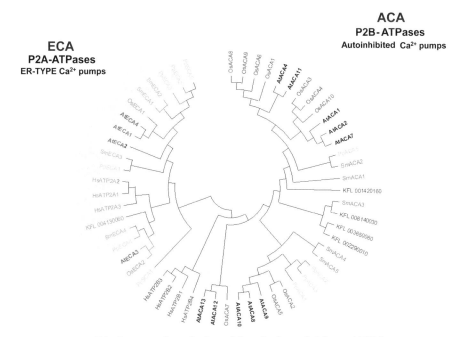

Phylogenetic relationship of plant ACA and ECA

Fig. 9.5 A phylogenetic tree depicting the members of ACA and ECA in plants. The unrooted Neighbor-Joining tree depicts the family members present in *Arabidopsis thaliana* (At), *Physcomitrela Patens* (Pp), *Selaginella moelendorfii* (Sm), *Klebsormidium nitens* (KFL), *Oryza sativa* (Os) and *Homo sapiens* (Hs). The sequences were taken from Pedersen et al. (2012), Singh et al. (2014) and UniProt. Tree constructed using MEGA10

been further implicated in pollen tube fertilization and development, plant development, gravity sensing, sucrose signaling, K^+ homeostasis, salicylic acid-mediated programmed cell death pathway, and plant immunity (Lucca and León 2012; Bushart et al. 2014; Iwano et al. 2014; Zhang et al. 2014; Costa et al. 2017; Demidchik et al. 2018). ECA1 is also responsible in generating long distance Ca^{2+} signal (Shkolnik et al. 2018).

Ca²⁺/Cation Antiporter (CaCA)

Similar to the P-type ATPases, the Ca^{2+}/cation antiporter (CaCA) is also a multi-member superfamily, and most members are associated with Ca^{2+} transport (Emery et al. 2012; Pittman and Hirschi 2016b). YRBG, Na^+/Ca^{2+} exchanger (NCX), Na^+/Ca^{2+}, K^+ exchanger (NCKX), cation Ca^{2+} exchanger (CCX) and H^+/Cation exchanger (CAX) are the major members of the CaCA superfamily (Emery et al. 2012). The YRBG are specific to bacteria and archaea (Emery et al. 2012). Similarly, the NCKX are also not present in land plants (although some algae show its presence) (Emery et al. 2012; Pittman and Hirschi 2016b). Typical NCX are also not present in land plants (again algae possess them), but Mg^{2+}/H^+ exchanger (MHX) (belonging to NCX family) are present in land plants (Shaul et al. 1999; Emery et al. 2012; Pittman and Hirschi 2016b). As the name suggests, MCX is not involved in Ca^{2+} transport, and are the exception in the CaCA superfamily (Shaul et al. 1999). A new member, Na^+/Ca^{2+} exchanger-like protein (NCL) were identified in plants that are more similar to CAX (based on phylogenetic analysis) rather than the NCX (Emery et al. 2012; Wang et al. 2012). Figure 9.6 shows the phylogenetic relationship between different members of the CaCA superfamily.

The CAX family can be considered to be plant specific family (exceptions fungi, protists, algae) and are absent from animals (mammals, insects, and nematodes) (Emery et al. 2012). The CCX are also majorly plant dominant (exceptions algae) but a single gene is present in human (Emery et al. 2012). There is another fundamental difference in working of the CaCA proteins in animal and plants. While the animals primarily use Na^+ as a counter ion (for the CaCA's antiport activity), plants use H^+ as counterion (Emery et al. 2012). The members of CaCA superfamily have the conserved structure of TM helices (that vary depending on the member) embedded into the membrane and posses two α-repeats regions for ion transport (Emery et al. 2012). Figure 9.7 shows the hypothetical membrane topology of the CaCA members present in plants (and involved in Ca^{2+} transport)- CAX, CCX and NCL. We discuss more on each group member in the subsequent sections.

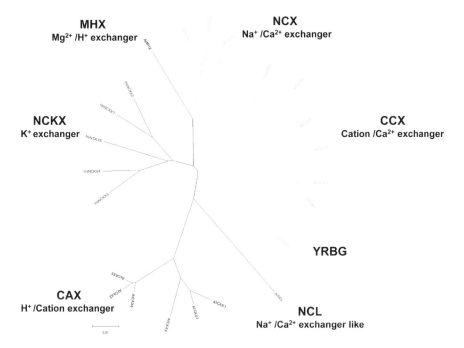

Fig. 9.6 A phylogenetic tree depicting the CaCA superfamily. An unrooted Neighbor-Joining tree depicts the members of the CaCA superfamily. Sequences are from *Escherichia coli* (YRBG), *Arabidopsis thaliana* (At), and *Homo sapiens* (Hs). The sequences were taken from UniProt and classification was done according to Emery et al. (2012), Pittman and Hirschi (2016b). Tree constructed using MEGA10

CAX

The CAX proteins were classified in three major phylogenetic clades based on their amino acid sequences- Type I (similar to *A. thaliana* CAX1), Type II (similar to *Saccharomyces cerevisiae* VNX1) and Type III (CAXs similar to *E. coli* ChaA) (Shigaki et al. 2006; Manohar et al. 2011). The Type I CAX are specific plant CAX and can be divided into another two clades TYPE IA and TYPE IB (Shigaki and Hirschi 2006), but the functional consequence of this classification is not clear (as most of the properties, like ion transport, from the two-member groups match) (Emery et al. 2012). As the plant CAX show broad range of ion specificity (besides Ca^{2+}), they are now referred to as "H$^+$/Cation exchanger" rather than the earlier nomenclature of "H$^+$/Ca^{2+} exchanger" (Emery et al. 2012). The plant CAX protein has 11 transmembrane domain, N- terminal regulatory or autoinhibitory domain

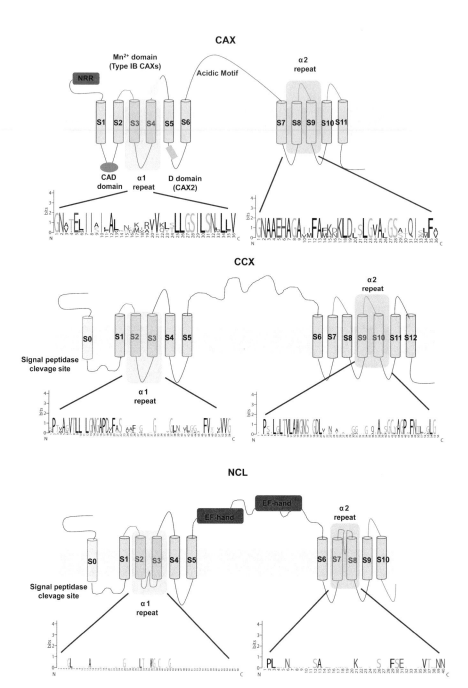

Fig. 9.7 Topological model of plant CAX, CCX and NCL. A hypothetical structure of the CAX, CCX and NCL antiporters with its domains. CAX is shown with its 11 TM domains and other domains as described in the text (figure adapted and modified from Yadav et al. 2012). The CCX with 12 TM helices, a signal peptidase cleavage site and a large loop connecting S5 and S6 (figure adapted and modified from Cai and Lytton 2004). The NCL has predicted 10 TM, two EF-hands and N-terminal signal peptide domain (Wang et al. 2012; Li et al. 2016). The N-terminal signal peptide is shown as a TM helix-like NCX proteins (Cai and Lytton 2004). Each group shows its consensus motif of alpha repeats. For CAX and CCX bona-fide members from Arabidopsis were used to generate the consensus. For NCL, TAIR database was used to search other NCL sequences using AtNCL amino acid sequence. Alignment logo generated using WebLogo

(NRR), Ca^{2+} domain (CaD), acidic motifs and cation selectivity filter, required for cation selection and transport. The NRR is for autoinhibition of CAX transport activity. It requires phosphorylation by CAX interacting Protein 1 (CXIP1) and CIPK24 for removal of the autoinhibition. The CaD domain provides Ca^{2+} and cation transport specificity to CAX. This is followed by the first α repeats (α1 also known as c-1 in literature) typical of CaCA family and this also play an important role in cation selectivity. The α2 (also known as c-2 in literature) (between TM S8 and S9) also performs a similar function (Cai and Lytton 2004; Shigaki and Hirschi 2006). The Mn^{2+} domain (also known as Type IB CAX or C-domain) is responsible for vacuolar metal/H^+ transport (Yadav et al. 2012). The D-domain is present in CAX2 and is responsible for pH sensitivity (Yadav et al. 2012). The loop region joining the TM6 and TM7 is called the "acidic motif" and it divides the CAX poly-peptide into two halves (Shigaki and Hirschi 2006). This region is different from other CaCA members as it is shorter in comparison (Cai and Lytton 2004; Shigaki and Hirschi 2006).

CAX polypeptides can form homomeric or heteromeric dimers or trimers (Demidchik et al. 2018). The autoinhibition is caused by the interaction of the N-terminal of two neighbors in the oligomer (Demidchik et al. 2018). The CAX are regulated by the NRR in a mechanism similar to the ACA, but the regulators in case of CAX are kinases (CIPK24 and CXIP1) and not CaM (Cheng et al. 2004a, b; Shigaki and Hirschi 2006). The CAX operate through a 'one H^+ in and one Ca^{2+} out' mechanism (Pittman and Hirschi 2016a).

As already mentioned several species posses CAX gene and they have been clas-sified into different phylogenetic clades according to their sequences. The study showed that CAX family has expanded in number in the plant kingdom. Like the Ca^{2+} transporters, CAX also has a presence at the PM and vacuole and are involved in regulating important plant physiological aspects (reviewed extensively in (Manohar et al. 2011; Yadav et al. 2012; Pittman and Hirschi 2016a).

CCX

The CCX was initially classified as members of the CAX family, but later phyloge-netic analysis classified them in a separate group with independent identity (due to their sequence similarity to mammalian NCKX6, the human CCX) (Shigaki et al. 2006; Emery et al. 2012). CCX has 12 TM helices and the same α repeats regions for ion transport (this region varies from the CAX). The loop joining the TM S5 and S6 is similar to the NCX and NCKX (Cai and Lytton 2004). The Ca^{2+} transport abil-ity of the CCX is debated as both reports from Arabidopsis CCX (CCX3, CCX4 and CCX5) indicate that they may not be involved in Ca^{2+} transport (Morris et al. 2008;

Zhang et al. 2011). CCX1 and CCX2 are involved in modulating Ca^{2+} dynamics in the cell (Mei et al. 2009; Corso et al. 2018). Similarly, the rice OsCCX2 mediates Ca^{2+} transport among other metals (Yadav et al. 2015; Hao et al. 2018). Compared to mammal and fungi, land plants have more CCX genes indicating the importance of CCX in plants (Emery et al. 2012).

NCL

The NCL was earlier classified as EF-CAX as they showed sequence similarity with CAX (Emery et al. 2012), but later they were classified into a new clade NCL (Wang et al. 2012; Pittman and Hirschi 2016b). The two EF-hands present in the long loop connecting TM5 and TM6 differentiate NCL from the other plant CaCA members (Wang et al. 2012). Their α repeats are also different from CAX (Emery et al. 2012). They can bind to Ca^{2+} and are vacuolar Na^+/Ca^{2+} exchanger (Wang et al. 2012; Li et al. 2016). Besides regulating Na^+/Ca^{2+} homeostasis, it is also involved in regulating auxin signaling and flowering time (Wang et al. 2012; Li et al. 2016). These genes are present in land plants as well as in algae (Emery et al. 2012).

Conclusion

The Ca^{2+} transporting members of the P-type ATPases and CaCA family of plants work to keep Ca^{2+} concentration in the cell at basal condition (explained in Fig. 9.8). These two transport elements work in to extrude Ca^{2+} from the cytosol and maintain the resting Ca^{2+} concentration in the cytosol. There are several other components modulating their activities (like CaM and Ca^{2+} regulated kinases (CDPKs or CIPKs). Together they form modules that regulate the extrusion machinery to maintain Ca^{2+} homeostasis.

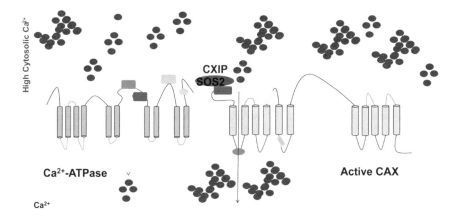

Transient Ca²⁺ rise in cytosol is mitigated majorly through CAX

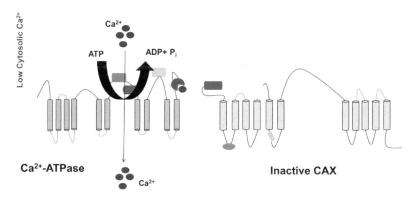

Under resting condition-Ca²⁺ removal majorly by Ca²⁺-ATPases

Fig. 9.8 A hypothetical model explaining the working of ATPase and CaCA in plants for maintaining basal Ca²⁺ levels in the cell. The basal Ca^{2+} concentration is maintained by synergistic action of ATPases (low capacity and high affinity for Ca^{2+}) and CAX (high capacity and low affinity for Ca^{2+}). The cytosolic Ca^{2+} transient is brought down by the CAX proteins, which are activated by phosphorylation of the NRR of CAX by CIPK24 or CXIP. They bring down the Ca^{2+} concentration to about 10 μM, after which the ATPases take over (due to the modulation by CaM) and extrude Ca^{2+} further to maintain the basal concentration around 100 nM in the cytosol. CAX are inactivated at low Ca^{2+} concentration. (Adapted and modified from Yadav et al. 2012)

References

I. Altshuler, J.J. Vaillant, S. Xu, M.E. Cristescu, The evolutionary history of Sarco(endo)plasmic Calcium ATPase (SERCA). PLoS One **7**, e52617 (2012)

K.B. Axelsen, M.G. Palmgren, Inventory of the superfamily of P-type ion pumps in Arabidopsis. Plant Physiol. **126**, 696–706 (2001)

Y. Boursiac, S.M. Lee, S. Romanowsky, R. Blank, C. Sladek, W.S. Chung, J.F. Harper, Disruption of the vacuolar calcium-ATPases in Arabidopsis results in the activation of a salicylic acid-dependent programmed cell death pathway. Plant Physiol. **154**, 1158–1171 (2010)

T.J. Bushart, A. Cannon, G. Clark, S.J. Roux, Structure and function of CrACA1, the major PM-type Ca2+-ATPase, expressed at the peak of the gravity-directed trans-cell calcium current in spores of the fern Ceratopteris richardii. Plant Biol. (Stuttg.) **16**(Suppl 1), 151–157 (2014)

X. Cai, J. Lytton, The cation/ca(2+) exchanger superfamily: Phylogenetic analysis and structural implications. Mol. Biol. Evol. **21**, 1692–1703 (2004)

N.H. Cheng, J.K. Pittman, J.K. Zhu, K.D. Hirschi, The protein kinase SOS2 activates the Arabidopsis H$^+$/Ca^{2+} antiporter CAX1 to integrate calcium transport and salt tolerance. J. Biol. Chem. **279**, 2922–2926 (2004a)

N.H. Cheng, J.Z. Liu, R.S. Nelson, K.D. Hirschi, Characterization of CXIP4, a novel Arabidopsis protein that activates the H+/Ca2+ antiporter, CAX1. FEBS Lett. **559**, 99–106 (2004b)

M. Corso, F.G. Doccula, J.R.F. de Melo, A. Costa, N. Verbruggen, Endoplasmic reticulum-localized CCX2 is required for osmotolerance by regulating ER and cytosolic Ca2+ dynamics in Arabidopsis. Proc. Natl. Acad. Sci. U. S. A. **115**, 3966–3971 (2018)

A. Costa, L. Luoni, C.A. Marrano, K. Hashimoto, P. Koster, S. Giacometti, M.I. De Michelis, J. Kudla, M.C. Bonza, Ca2+−dependent phosphoregulation of the plasma membrane Ca2+-ATPase ACA8 modulates stimulus-induced calcium signatures. J. Exp. Bot. **68**, 3215–3230 (2017)

V. Demidchik, S. Shabala, S. Isayenkov, T.A. Cuin, I. Pottosin, Calcium transport across plant membranes: Mechanisms and functions. New Phytol. **220**, 49–69 (2018)

L. Emery, S. Whelan, K.D. Hirschi, J.K. Pittman, Protein phylogenetic analysis of Ca2+/cation Antiporters and insights into their evolution in plants. Front. Plant Sci. **3**, 1 (2012)

N. Frei dit Frey, M. Mbengue, M. Kwaaitaal, L. Nitsch, D. Altenbach, H. Häweker, R. Lozano-Duran, M.F. Njo, T. Beeckman, B. Huettel, J.W. Borst, R. Panstruga, S. Robatzek, Plasma membrane calcium ATPases are important components of receptor-mediated signaling in plant immune responses and development. Plant Physiol. **159**, 798–809 (2012)

S. Giacometti, C.A. Marrano, M.C. Bonza, L. Luoni, M. Limonta, M.I. De Michelis, Phosphorylation of serine residues in the N-terminus modulates the activity of ACA8, a plasma membrane Ca2+-ATPase of Arabidopsis thaliana. J. Exp. Bot. **63**, 1215–1224 (2012)

X. Hao, M. Zeng, J. Wang, Z. Zeng, J. Dai, Z. Xie, Y. Yang, L. Tian, L. Chen, D. Li, A node-expressed transporter OsCCX2 is involved in grain cadmium accumulation of Rice. Front. Plant Sci. **9**, 476 (2018)

B. Hong, A. Ichida, Y. Wang, J. Scott Gens, B.G. Pickard, J.F. Harper, Identification of a Calmodulin-regulated Ca2+-ATPase in the endoplasmic reticulum. Plant Physiol. **119**, 1165–1176 (1999)

L. Huang, T. Berkelman, A.E. Franklin, N.E. Hoffman, Characterization of a gene encoding a Ca(2+)-ATPase-like protein in the plastid envelope. Proc. Natl. Acad. Sci. U. S. A. **90**, 10066–10070 (1993)

K.M. Huda, M.S. Banu, R. Tuteja, N. Tuteja, Global calcium transducer P-type Ca^{2+}-ATPases open new avenues for agriculture by regulating stress signalling. J. Exp. Bot. **64**, 3099–3109 (2013)

I. Hwang, H. Sze, J.F. Harper, A calcium-dependent protein kinase can inhibit a calmodulin-stimulated Ca2+ pump (ACA2) located in the endoplasmic reticulum of Arabidopsis. Proc. Natl. Acad. Sci. U. S. A. **97**, 6224–6229 (2000)

M. Iwano, M. Igarashi, Y. Tarutani, P. Kaothien-Nakayama, H. Nakayama, H. Moriyama, R. Yakabe, T. Entani, H. Shimosato-Asano, M. Ueki, G. Tamiya, S. Takayama, A pollen coat–inducible autoinhibited Ca2+-ATPase expressed in stigmatic papilla cells is required for compatible pollination in the Brassicaceae. Plant Cell **26**, 636–649 (2014)

P. Li, G. Zhang, N. Gonzales, Y. Guo, H. Hu, S. Park, J. Zhao, Ca(2+) -regulated and diurnal rhythm-regulated Na(+)/Ca(2+) exchanger AtNCL affects flowering time and auxin signalling in Arabidopsis. Plant Cell Environ. **39**, 377–392 (2016)

N. Lucca, G. León, Arabidopsis ACA7, encoding a putative auto-regulated Ca(2+)-ATPase, is required for normal pollen development. Plant Cell Rep. **31**, 651–659 (2012)

M. Manohar, T. Shigaki, K.D. Hirschi, Plant cation/H+ exchangers (CAXs): Biological functions and genetic manipulations. Plant Biol. (Stuttg.) **13**, 561–569 (2011)

H. Mei, N.H. Cheng, J. Zhao, S. Park, R.A. Escareno, J.K. Pittman, K.D. Hirschi, Root development under metal stress in Arabidopsis thaliana requires the H+/cation antiporter CAX4. New Phytol. **183**, 95–105 (2009)

R.F. Mills, M.L. Doherty, R.L. López-Marqués, T. Weimar, P. Dupree, M.G. Palmgren, J.K. Pittman, L.E. Williams, ECA3, a Golgi-localized P2A-type ATPase, plays a crucial role in manganese nutrition in Arabidopsis. Plant Physiol. **146**, 116–128 (2008)

J. Morris, H. Tian, S. Park, C.S. Sreevidya, J.M. Ward, K.D. Hirschi, AtCCX3 is an Arabidopsis endomembrane H+-dependent K+ transporter. Plant Physiol. **148**, 1474–1486 (2008)

J.P. Morth, B.P. Pedersen, M.J. Buch-Pedersen, J.P. Andersen, B. Vilsen, M.G. Palmgren, P. Nissen, A structural overview of the plasma membrane Na+,K+-ATPase and H+-ATPase ion pumps. Nat. Rev. Mol. Cell Biol. **12**, 60–70 (2011)

M.G. Palmgren, P. Nissen, P-type ATPases. Annu. Rev. Biophys. **40**, 243–266 (2011)

C.N.S. Pedersen, K.B. Axelsen, J.F. Harper, M.G. Palmgren, Evolution of plant P-type ATPases. Front. Plant Sci. **3**, 31 (2012)

J.K. Pittman, K.D. Hirschi, Phylogenetic analysis and protein structure modelling identifies distinct Ca2+/Cation antiporters and conservation of gene family structure within Arabidopsis and rice species. Rice (N Y) **9**, 3 (2016a)

J.K. Pittman, K.D. Hirschi, CAX-ing a wide net: Cation/H(+) transporters in metal remediation and abiotic stress signalling. Plant Biol. **18**, 741–749 (2016b)

M. Schiott, S.M. Romanowsky, L. Baekgaard, M.K. Jakobsen, M.G. Palmgren, J.F. Harper, A plant plasma membrane Ca2+ pump is required for normal pollen tube growth and fertilization. Proc. Natl. Acad. Sci. U. S. A. **101**, 9502–9507 (2004)

O. Shaul, D.W. Hilgemann, J. de-Almeida-Engler, M. Van Montagu, D. Inz, G. Galili, Cloning and characterization of a novel Mg2+/H+ exchanger. EMBO J. **18**, 3973–3980 (1999)

T. Shigaki, K.D. Hirschi, Diverse functions and molecular properties emerging for CAX cation/H+ exchangers in plants. Plant Biol. (Stuttg.) **8**, 419–429 (2006)

T. Shigaki, I. Rees, L. Nakhleh, K.D. Hirschi, Identification of three distinct phylogenetic groups of CAX cation/proton antiporters. J. Mol. Evol. **63**, 815–825 (2006)

D. Shkolnik, R. Nuriel, M.C. Bonza, A. Costa, H. Fromm, MIZ1 regulates ECA1 to generate a slow, long-distance phloem-transmitted Ca2+ signal essential for root water tracking in Arabidopsis. Proc. Natl. Acad. Sci. U. S. A. **115**, 8031–8036 (2018)

A. Singh, P. Kanwar, A.K. Yadav, M. Mishra, S.K. Jha, V. Baranwal, A. Pandey, S. Kapoor, A.K. Tyagi, G.K. Pandey, Genome-wide expressional and functional analysis of calcium transport elements during abiotic stress and development in rice. FEBS J. **281**, 894–915 (2014)

M.D. Thever, M.H. Saier, Bioinformatic characterization of P-type ATPases encoded within the fully sequenced genomes of 26 eukaryotes. J. Membr. Biol. **229**, 115–130 (2009)

L.J. Thomson, T. Xing, J.L. Hall, L.E. Williams, Investigation of the calcium-transporting ATPases at the endoplasmic reticulum and plasma membrane of red beet (Beta vulgaris). Plant Physiol. **102**, 553–564 (1993)

H. Tidow, L.R. Poulsen, A. Andreeva, M. Knudsen, K.L. Hein, C. Wiuf, M.G. Palmgren, P. Nissen, A bimodular mechanism of calcium control in eukaryotes. Nature **491**, 468–472 (2012)

P. Wang, Z. Li, J. Wei, Z. Zhao, D. Sun, S. Cui, A Na+/Ca2+ exchanger-like protein (AtNCL) involved in salt stress in Arabidopsis. J. Biol. Chem. **287**, 44062–44070 (2012)

L.E. Wimmers, N.N. Ewing, A.B. Bennett, Higher plant Ca2+-ATPase: Primary structure and regulation of mRNA abundance by salt. Proc. Natl. Acad. Sci. U. S. A. **89**, 9205–9209 (1992)

Z. Wu, F. Liang, B. Hong, J.C. Young, M.R. Sussman, J.F. Harper, H. Sze, An endoplasmic reticulum-bound Ca2+/Mn2+ pump, ECA1, supports plant growth and confers tolerance to Mn2+ stress. Plant Physiol. **130**, 128–137 (2002)

A.K. Yadav, A. Pandey, G.K. Pandey, Calcium Homeostasis: Role of CAXs Transporters in Plant Signaling. Plant Stress. **6**, 60–69 (2012)

A.K. Yadav, A. Shankar, S.K. Jha, P. Kanwar, A. Pandey, G.K. Pandey, A rice tonoplastic calcium exchanger, OsCCX2 mediates Ca^{2+}/cation transport in yeast. Sci. Rep. **5**, 17117 (2015)

X. Zhang, M. Zhang, T. Takano, S. Liu, Characterization of an AtCCX5 gene from Arabidopsis thaliana that involves in high-affinity K$^+$ uptake and Na$^+$ transport in yeast. Biochem. Biophys. Res. Commun. **414**, 96–100 (2011)

J. Zhang, X. Zhang, R. Wang, W. Li, The plasma membrane-localised Ca(2+)-ATPase ACA8 plays a role in sucrose signalling involved in early seedling development in Arabidopsis. Plant Cell Rep. **33**, 755–766 (2014)

Chapter 10
Tools for Analysing Ca²⁺ Transport Elements and Future Perspectives

Contents

Introduction

The study of plant Ca^{2+} signaling and homeostasis in general and the Ca^{2+} transport elements (CTE) have helped us understand the basic biology of plant signal transduction (via Ca^{2+} signaling pathway). To understand the functional role of different Ca^{2+} channels and transporters, the importance of the several tools and techniques such as electrophysiology, complementation assay in yeast and other heterologous systems, *in vivo* Ca^{2+} measurement in the cell, genetic based approaches such as loss-of-function mutation and gain-of-function analysis, cell biological analysis have been extensively used by researchers. Researchers have also increasingly used high throughput 'omics' approaches (e.g., transcriptomics and proteomics to determine transcriptional regulatory controls and post translational modifications (PTMs)). The study of Ca^{2+} transport elements involves some specialized techniques that we discuss in the subsequent sections. We also briefly discuss the role of CTE in long-distance signaling and plant intelligence. Additionally, we also complete this write up with some important questions for future research in the field of Ca^{2+} transport biology.

© The Editor(s) (if applicable) and The Author(s), under exclusive license to
Springer Nature Switzerland AG 2021
G. K. Pandey, S. K. Sanyal, *Functional Dissection of Calcium Homeostasis and Transport Machinery in Plants*, SpringerBriefs in Plant Science,
https://doi.org/10.1007/978-3-030-58502-0_10

Electrophysiology

We have mentioned several times during our discussions on the CTE about electrophysiological validation for channel conductance. The ion channels form a gateway within the membrane through which ions are passively exchanged depending on the difference in concentration (ions flowing from high to low concentration zones) and in case of the electrically charged membrane, electrical charges, as well as concentration difference, modulate the flow of ions (Demidchik et al. 2006; Demidchik 2012). We have already mentioned the different properties and types of Ca^{2+} ion channels in Chap. 5. Although different electrophysiological techniques allow characterizing the channel function *in vivo*, two main electrophysiological approaches such as extracellular and intracellular recordings is involved extensively in characterising plant ion channels (Demidchik 2012).

The Fields-potential is measured using an electrode in the extracellular medium. This is a non-destructive technique and has been used for examination of electrical responses of roots and leaves (Demidchik 2012). An advanced method is the Microelectrode Ion Flux Estimation (MIFE®) and is used extensively in extracellular measurements (Demidchik 2012; Shabala and Bose 2012; Shabala et al. 2012). Here electrodes are used to measure ionic fluxes in the root cells. The MIFE® technique is also non-destructive and has several other advantages (like selectivity to ion fluxes, measurement of several ion fluxes simultaneously among others). However, there are certain limitations to both these techniques. The former provides only "change in electric field potential data" which cannot be directly interpreted as ion flux. The later does not measure the fluxes by internal tissues and is also not suited for measurement of channel kinetics. These factors limit the use of these techniques (Demidchik 2012).

The intracellular techniques, as the name suggests, are invasive. There are two ways, in which the conductance can be measured-using a two-electrode voltage clamp (TEVC) and a patch-clamp technique (Demidchik 2012). TEVC can measure the ion conductance at a given voltage and does not damage the cell. However, there are certain problems (like problems with impaling, cell size among others) that limit its use in characterizing channels of higher plants (Demidchik 2012). So instead of using intact roots, cation channels are expressed in the *Xenopus laevis* oocytes and characterized using the TEVC technique (Dreyer et al. 1999; Miller and Zhou 2000; Leng et al. 2002; Mori et al. 2018; Pan et al. 2019; Tian et al. 2019). The patch-clamp technique replaced the TEVC as it used protoplasts and allowed ion flux measurement of a large number of plants (Schroeder et al. 1984; Demidchik 2012; Hedrich 2012; Hedrich et al. 2012). There are different modes of patch-clamp measurements- cell attached, whole-cell, inside-out and outside-out modes. Each mode has its own sets of advantages and disadvantages (reviewed extensively in (Demidchik 2012)). The advantage that patch-clamp technique provides in allowing measuring protoplast of any plant tissue comes with an attached disadvantage- the treatment to remove the cell wall (to isolate protoplast), can also affect the native

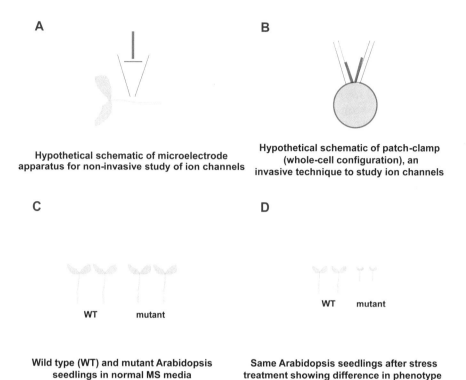

A Hypothetical schematic of microelectrode apparatus for non-invasive study of ion channels

B Hypothetical schematic of patch-clamp (whole-cell configuration), an invasive technique to study ion channels

C Wild type (WT) and mutant Arabidopsis seedlings in normal MS media

WT   mutant

D Same Arabidopsis seedlings after stress treatment showing difference in phenotype

WT   mutant

Fig. 10.1 **Hypothetical schematic model of electrophysiological and plant phenotyping techniques**. (**a**) A hypothetical schematic showing the MIFE® technique used for non-invasive characterization of CTE. (**b**) Whole cell configuration of patch-clamp technique for characterization of CTE. (Figure (**a**) and (**b**) adapted from Demidchik 2012). Phenotyping Arabidopsis seedling by (**c**) plating wild type (WT) and mutant for CTE (mutation generated by different methodologies described in text) in normal media. (**d**) Same seedlings are subjected to stressed and a phenotype difference is observed (if any). In the figure, a hypersensitive mutant is shown. The mutant may also have tolerant (to stress) phenotype and perform better than the WT. The phenotype is used to assess the *in planta* function of the CTE

channels (Demidchik 2012). Figure 10.1 provides a brief summary of the invasive and non-invasive techniques used in electrophysiology.

The planar lipid bilayer technique that has been used in the past for electrophysiological analysis of cation channels (White 1993, 1994, 1998; White and Tester 1994). This technique provides the advantage of monitoring the channel behavior at the single-molecule level, thus identifying factors required for channel activity (Zakharian 2013). This has a disadvantage- compared to patch clamps, in generating more "noise" and channels studied by this technique shows slow voltage response times (Zakharian 2013).

Expression in a Heterologous System

Another method for characterization of CTE is by expressing them in a heterologous system, which is devoid of endogenous Ca^{2+} export mechanism. Again the heterologous system allows characterizing the CTE without hindrance that would have been faced had they been examined in the native system (Wang 2012). We have already discussed the TEVC and the *X. laevis* oocytes, which serve as a very good example of heterologous expression and investigation of ion channels. Besides oocytes, electrophysiological experiments can be performed in Human Embryonic Kidney (HEK) cells (Christopher et al. 2007; Hamada et al. 2012; Vincill et al. 2012, 2013; Guo et al. 2016), Chinese Hamster Ovary (CHO) cells (Hou et al. 2014) and *E. coli* spheroplasts (Schlegel and Haswell 2020). In this section, we discuss the yeast systems used in the characterization of plant transport elements. *Saccharomyces cerevisiae* has proven to be an important tool for investigating and characterizing plant ion transport elements in general and that belonging to Ca^{2+} transport in particular (Sze et al. 2000; Wang 2012). Creating loss-of-function mutants of yeast lacking important transport elements in the cell have yielded important mutant collection from the perspective of transport element characterization. There are two Ca^{2+} pumps- PMR1 at the Golgi and PMC1 at the vacuolar membrane. A triple mutant of wild type yeast lacking both pumps and Calcineurin B1 (because double knockout *pmc1pmr1* not viable, therefore, *pmr1pmc1cnb1* is viable-triple mutant named as K616) has proven to be a potent tool to characterize plant Ca^{2+} ATPases and CCX (Sze et al. 2000). Another mutant K667 (lacking *PMC1*, *CNB1* and *VCX1* (an important Ca^{2+}/H^+ exchanger in yeast vacuole)) has also been used to characterize CTE (Edmond et al. 2009; Yadav et al. 2015). We had talked about the Mid1-complementing channel (MCA) in Chap. 8, which were identified based on complementing the *mid1* yeast mutant (lacking the MID1 Ca^{2+} permeable channel) (Ali et al. 2005). These mutants with their location in yeast cell is depicted in Fig. 10.2.

Ca²⁺ Sensors for *in vivo* Ca²⁺ Measurements

A parallel approach to study CTE is to use Ca^{2+} sensors (we categorize synthetic dyes as well as genetically encoded sensors in this class in this write-up) and fluorescence microscopes. They provide an alternate experimental technique to study the Ca^{2+} dynamics *in planta,* and we are describing the different classes of Ca^{2+} sensors and dyes subsequently. Initially, chemical indicators (dyes) were used to measure the Ca^{2+} dynamics in the cell. The first candidates were synthetic dyes, like azo dyes, murexide and chlortetracycline were used as indicators for Ca^{2+} detection (Kanchiswamy et al. 2014). Non-ratiometric dyes (single wavelength probes) with single excitation spectrum (e.g., Fluo, Rhod and Calcium Green-1) were used due to the ease of quantification (Kanchiswamy et al. 2014). Ratiometric dyes (e.g., Fura-2 and Indo-1) vary their spectrum according to free Ca^{2+} concentration

A

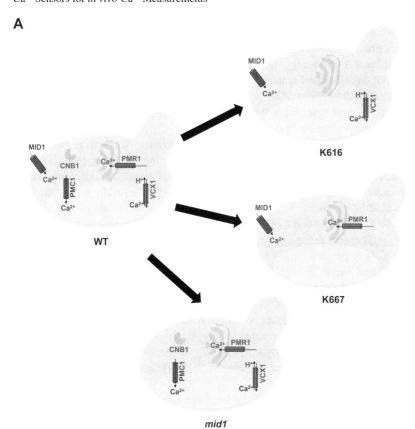

Different yeast mutants used for characterization of Ca²⁺ transport elements

B

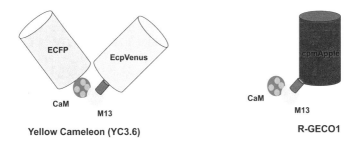

Two FP GECI YC3.6 and Single FP GECI R-GECO1

Fig. 10.2 Hypothetical model of yeast heterologous system and genetically encoded Ca²⁺ indicators. (**a**) Hypothetical model showing the mutant yeast strains used for characterization of Ca²⁺ transport elements. The wild type (WT) yeast with important yeast transport elements and CNB1 is shown. The different mutants K616 (*pmr1pmc1cnb1*), K667 (*pmc1vcx1cnb1*), and *mid1* mutant. (Figure adapted and modified from Sze et al. 2000). (**b**) Two important GECIs used extensively in plant research are shown. The YC3.6 with two fluorescent proteins ECFP and cpVenus. The R-GECO1 is a single fluorescent protein mApple. Both the GECIs have CaM for Ca²⁺ binding and M13 connector peptide. (Figure adapted and modified from Costa and Kudla 2015)

(Kanchiswamy et al. 2014). These chemical indicators are easily available (commercially), provide a very strong signal for analysis (through a fluorescence microscope) and have a wide range of Ca^{2+} binding affinity. Another factor is the fast generation of experimental material with the use of dyes that goes in its favor. But there are several drawbacks of these dyes- they can cause photo-damage and their signal decreases over time. Moreover, they lack the desired accuracy to study single cell dynamic (Kanchiswamy et al. 2014).

These shortcomings led scientists to develop alternate ways to perform Ca^{2+} imaging *in vivo*. The development of this particular tool owes its genesis to the Aequorin protein, which can bind to Ca^{2+} and emit luminescence (after incubation with coelenterazine co-factor) (Costa and Kudla 2015). Aequorin was fused with other fluorescent proteins (FP) (like GFP, mVENUS etc) to create Genetically Encoded Ca^{2+} Indicators (GECI). One the most common structure of a GECI is an FP (or bioluminescent protein) is attached to a Ca^{2+} sensor (like Calmodulin (CaM)) and CaM binding peptide (M13)-representation of a typical single FP GECI (Pérez Koldenkova and Nagai 2013; Costa and Kudla 2015). In another typical architecture, the CaM-M13 (or variant) is attached between two FP (or two different proteins) generating a two FP-GECI (Pérez Koldenkova and Nagai 2013; Costa and Kudla 2015). Ca^{2+} binding causes structural rearrangements, which would either change the fluorescence due to change in absorption resulting in the change of emitted fluorescence (in case of single FP GECI) or use Förster Resonance Energy Transfer (FRET) (in case of two FP based GECI) (Pérez Koldenkova and Nagai 2013; Costa and Kudla 2015).

Several other forms of GECIs have been developed over the years and have allowed better measurement of Ca^{2+} dynamics at the cellular level. The previously mentioned Aequorin-GFP based GECI functions on the principle of bioluminescent resonance energy transfer (BRET) (Pérez Koldenkova and Nagai 2013). Based on the same principles, BRAC (a luciferase-based GECI) and Nano-lantern GECIs were developed. In BRAC the RLuc8-CaM-M13-fluorescent protein Venus form the core body of the GECI (Pérez Koldenkova and Nagai 2013). The Nano-lantern is a modified BRAC, with the CaM-M13 inserted in a split Rluc8. These sensors report a change in Ca^{2+} dynamics in the non-ratiometric manner (Pérez Koldenkova and Nagai 2013).

Besides the GECIs based on bioluminescence, there is another set of GECI based on fluorescence. The FRET-based GECIs can be further classified into single FP based GECI (SFPG) and two FP based GECI (TFPG) (Pérez Koldenkova and Nagai 2013; Costa and Kudla 2015). The CaM-M13 help in the Ca^{2+} binding part in the SPFG, and Ca^{2+} binding changes the fluorescence intensity and these are also non-ratiometric (Pérez Koldenkova and Nagai 2013). The GCaMP, GECO (genetically encoded Ca^{2+} indicators for optical imaging), Pericams, Camgroos, CatchER are examples of SFPG. The Camgroos only have CaM between two halves of a GFP molecule. The CatcherER are modified to carry a Ca^{2+} binding motif within the sensor (it does not have a separate CaM-M13 motif) (Pérez Koldenkova and Nagai 2013). The TFPG are based on FRET. Two FP of the SFPG come closer together when the probe binds Ca^{2+} and as a result of FRET, Ca^{2+} binding dynamics can be measured and these are ratiometric indicators. Cameleon and TN are examples of SFPG (Pérez Koldenkova and Nagai 2013). Several of these GECIs (majorly Cameleon and GECO) have been used for analyzing the Ca^{2+} dynamics to understand the function of CTE (Iwano et al. 2009; Krebs et al. 2012; Loro et al. 2012, 2016; Bonza et al. 2013; Choi et al. 2014; Ngo et al. 2014; Costa et al. 2017; Teardo et al. 2017). The Cameleon and GECO are represented in Fig. 10.2.

Functional Genomics Analysis

The identification and characterization of CTE by the above techniques assign Ca^{2+} transport to a particular protein. Scientists also employ other ways to identify the CTEs. The genome sequencing initiatives taken for different plant species have helped in the identification of novel CTEs in plant species. The "genome wide" studies are fast and efficient ways to identify and classify the CTEs and can be later verified using other experimental tools (Singh et al. 2014). The high throughput transcriptomics (e.g., microarray) approach has also led to the identification of CTEs involved in different physiological responses (Shankar et al. 2014). The analysis of a CTE is complete in a sense when the data from the above techniques is integrated to its functional role *in planta*. For this, knockout mutants (T-DNA insertion mutants, CRISPR-Cas9 mutants among others), knockdown (RNAi), overexpression lines of respective CTE are used for functional analysis (briefly shown in Fig. 10.1). Besides, forward genetic screens are extensively used to identify novel CTE involved in certain plant physiological phenomenon (Clough et al. 2000; Charpentier 2018). Another high throughput approach to identify PTMs is through chromatography followed by mass spectrometry of plant samples (nano-liquid chromatography tandem mass-spectrometry (nanoLC-MS/MS)). This approach has allowed the identification of important *in vivo* phosphorylation sites in CTEs (Nühse et al. 2003, 2007; Benschop et al. 2007; Niittyla et al. 2007; Chen et al. 2010). In the case where the effect of PTM need to be examined, the mutagenized version of CTE is expressed in the plants to determine the *in planta* effect (mostly through morphological or physiological effect).

Ca²⁺ in Memory and Long-Distance Signal Transport

Very recently a report from Gilroy laboratory has eloquently demonstrated that in response to the wounding, the signal can be perceived and transmitted throughout the plant body with the help of glutamate receptors (GLR) (Toyota et al. 2018). In fact plant long-distance signaling is thought to be very closely connected to crosstalk between the ROS and Ca^{2+} signaling pathway (Choi et al. 2016). The Ca^{2+} signaling pathway is related to another hypothetical phenomenon in plants-memory (Thellier and Luttge 2013; Trewavas 2016). Many plant biologists have put forward this concept and strongly believe that Ca^{2+} plays a significant role in plant memory (reviewed in (van Loon 2016; Sanyal et al. 2019)). As experiments have proven the role of CTE in long-distance signaling, it can also be hypothesized that they play a similar role in modulating the plant memory.

Future Perspective

The key questions listed below, are some points-to-ponder for Ca^{2+} signaling research in general and CTE research in particular.

1. Plants have gone through long evolutionary course to adapt themselves to the terrestrial environment. With the availability of the newly sequenced genome, we can have a better picture of the evolution of the CTE. This should help us know the changes that were incorporated during evolution and better understand the regulation of Ca^{2+} signaling and CTE.

2. Also, it is important to know the similarities and differences between the Ca^{2+} signaling events (homeostasis, CTE, and others) between the current crop species and their wild type relatives. This is important for the generation of better and future-ready crops.

3. The components involved in Ca^{2+} signaling events may vary between the cell types of the same plant, with the overall architecture of the homeostasis events staying the same. Deciphering them is another important task to better understand how the cells co-ordinate to make the plant functional entity.

4. We know the major decoders of Ca^{2+} signaling pathways, however, there are possibilities that many decoders are yet to be discovered and might be working in organelle-specific manner. Investigating them will enhance our knowledge of organellar Ca^{2+} signal decoding and tell more about the contribution of organelles in the plant physiology (in terms of stress perception and plant development).

5. We do know about the regulation of the CTE, but there may be more players involved in the regulation of the CTE. Also, there is a possibility of more complex regulation by the interplay of more than one protein and ion. The more their regulation is investigated, the better will be our understanding of Ca^{2+} homeostasis.

6. Some pathways that crosstalk with Ca^{2+} signaling pathways are well established. But more information is needed as there are possibilities that many other pathways can crosstalk and modulate the CTEs as well as Ca^{2+} signaling. The ROS-Ca^{2+} crosstalk and ABA-Ca^{2+} crosstalk have become a paradigm in plant physiology. Again this information will help better understand plant biology and possibly lead to design better and future-ready crops.

7. The role of post-translational modifications (PTMs) is very important in modulating any protein and CTEs are no exception to this fact. A comprehensive map

of the CTE PTMs is another important field that will need significant effort and will pave the way for a better understanding of the regulation of CTE.

8. The molecular identification of the CTE talked about in Chap. 5 is necessary and will solve a long-standing mystery in the field of Ca^{2+} signaling.

References

R. Ali, R.E. Zielinski, G.A. Berkowitz, Expression of plant cyclic nucleotide-gated cation channels in yeast. J. Exp. Bot. **57**, 125–138 (2005)

J.J. Benschop, S. Mohammed, M. O'Flaherty, A.J. Heck, M. Slijper, F.L. Menke, Quantitative phosphoproteomics of early elicitor signaling in Arabidopsis. Mol. Cell. Proteomics **6**, 1198–1214 (2007)

M.C. Bonza, G. Loro, S. Behera, A. Wong, J. Kudla, A. Costa, Analyses of Ca2+ accumulation and dynamics in the endoplasmic reticulum of Arabidopsis root cells using a genetically encoded Cameleon sensor. Plant Physiol. **163**, 1230–1241 (2013)

M. Charpentier, Calcium signals in the plant nucleus: Origin and function. J. Exp. Bot **69**, 4165–4173 (2018)

Y. Chen, W. Hoehenwarter, W. Weckwerth, Comparative analysis of phytohormone-responsive phosphoproteins in Arabidopsis thaliana using TiO2-phosphopeptide enrichment and mass accuracy precursor alignment. Plant J. **63**, 1–17 (2010)

W.G. Choi, M. Toyota, S.H. Kim, R. Hilleary, S. Gilroy, Salt stress-induced Ca2+ waves are associated with rapid, long-distance root-to-shoot signaling in plants. Proc. Natl. Acad. Sci. U. S. A. **111**, 6497–6502 (2014)

W.G. Choi, R. Hilleary, S.J. Swanson, S.H. Kim, S. Gilroy, Rapid, long-distance electrical and calcium signaling in plants. Annu. Rev. Plant Biol. **67**, 287–307 (2016)

D.A. Christopher, T. Borsics, C.Y. Yuen, W. Ullmer, C. Andème-Ondzighi, M.A. Andres, B.H. Kang, L.A. Staehelin, The cyclic nucleotide gated cation channel AtCNGC10 traffics from the ER via Golgi vesicles to the plasma membrane of Arabidopsis root and leaf cells. BMC Plant Biol. **7**, 48 (2007)

S.J. Clough, K.A. Fengler, I. Yu, B. Lippok, R.K. Smith, A.F. Bent, The Arabidopsis dnd1 "defense, no death" gene encodes a mutated cyclic nucleotide-gated ion channel. Proc. Natl. Acad. Sci. U. S. A. **97**, 9323–9328 (2000)

A. Costa, J. Kudla, Colorful insights: Advances in imaging drive novel breakthroughs in Ca2+ signaling. Mol. Plant **8**, 352–355 (2015)

A. Costa, L. Luoni, C.A. Marrano, K. Hashimoto, P. Koster, S. Giacometti, M.I. De Michelis, J. Kudla, M.C. Bonza, Ca2+-dependent phosphoregulation of the plasma membrane Ca2+-ATPase ACA8 modulates stimulus-induced calcium signatures. J. Exp. Bot. **68**, 3215–3230 (2017)

V. Demidchik, Characterisation of root plasma membrane Ca^{2+}-permeable cation channels: Techniques and basic concepts, in *Plant Electrophysiology*, ed. by A. Volkov, (Springer, Berlin/Heidelberg, 2012), pp. 339–369

V. Demidchik, A. Sokolik, V. Yurin, Electrophysiological characterization of plant cation channels, in *Plant Electrophysiology*, ed. by A. G. Volkov, (Springer, Berlin/Heidelberg, 2006), pp. 173–185

I. Dreyer, C. Horeau, G. Lemaillet, S. Zimmermann, D.R. Bush, A. Rodríguez-Navarro, D.P. Schachtman, E.P. Spalding, H. Sentenac, R.F. Gaber, Identification and characterization of plant transporters. J. Exp. Bot. **50**, 1073–1087 (1999)

C. Edmond, T. Shigaki, S. Ewert, M.D. Nelson, J.M. Connorton, V. Chalova, Z. Noordally, J.K. Pittman, Comparative analysis of CAX2-like cation transporters indicates functional and regulatory diversity. Biochem. J. **418**, 145–154 (2009)

J. Guo, W. Zeng, Q. Chen, C. Lee, L. Chen, Y. Yang, C. Cang, D. Ren, Y. Jiang, Structure of the voltage-gated two-pore channel TPC1 from Arabidopsis thaliana. Nature **531**, 196–201 (2016)

H. Hamada, T. Kurusu, E. Okuma, H. Nokajima, M. Kiyoduka, T. Koyano, Y. Sugiyama, K. Okada, J. Koga, H. Saji, A. Miyao, H. Hirochika, H. Yamane, Y. Murata, K. Kuchitsu, Regulation of a Proteinaceous elicitor-induced Ca2+ influx and production of Phytoalexins by a putative voltage-gated Cation Channel, OsTPC1, in cultured Rice cells. J. Biol. Chem. **287**, 9931–9939 (2012)

R. Hedrich, Ion channels in plants. Physiol. Rev. **92**, 1777–1811 (2012)

R. Hedrich, D. Becker, D. Geiger, I. Marten, M. Roelfsema, Role of ion channels in plants, in *Patch Clamp Techniques. From Beginning to Advanced Protocols*, ed. by Y. Okada, (Springer, New York, 2012)

C. Hou, W. Tian, T. Kleist, K. He, V. Garcia, F. Bai, Y. Hao, S. Luan, L. Li, DUF221 proteins are a family of osmosensitive calcium-permeable cation channels conserved across eukaryotes. Cell Res. **24**, 632–635 (2014)

M. Iwano, T. Entani, H. Shiba, M. Kakita, T. Nagai, H. Mizuno, A. Miyawaki, T. Shoji, K. Kubo, A. Isogai, S. Takayama, Fine-tuning of the cytoplasmic Ca2+ concentration is essential for pollen tube growth. Plant Physiol. **150**, 1322–1334 (2009)

C.N. Kanchiswamy, M. Malnoy, A. Occhipinti, M.E. Maffei, Calcium imaging perspectives in plants. Int. J. Mol. Sci. **15**, 3842–3859 (2014)

M. Krebs, K. Held, A. Binder, K. Hashimoto, G. Den Herder, M. Parniske, J. Kudla, K. Schumacher, FRET-based genetically encoded sensors allow high-resolution live cell imaging of Ca²⁺ dynamics. Plant J. **69**, 181–192 (2012)

Q. Leng, R.W. Mercier, B.G. Hua, H. Fromm, G.A. Berkowitz, Electrophysiological analysis of cloned cyclic nucleotide-gated ion channels. Plant Physiol. **128**, 400–410 (2002)

G. Loro, I. Drago, T. Pozzan, F.L. Schiavo, M. Zottini, A. Costa, Targeting of Cameleons to various subcellular compartments reveals a strict cytoplasmic/mitochondrial Ca²⁺ handling relationship in plant cells. Plant J. **71**, 1–13 (2012)

G. Loro, S. Wagner, F.G. Doccula, S. Behera, S. Weinl, J. Kudla, M. Schwarzlander, A. Costa, M. Zottini, Chloroplast-specific in vivo Ca2+ imaging using yellow Cameleon fluorescent protein sensors reveals organelle-autonomous Ca2+ signatures in the Stroma. Plant Physiol. **171**, 2317–2330 (2016)

A.J. Miller, J.J. Zhou, Xenopus oocytes as an expression system for plant transporters. Biochim. Biophys. Acta **1465**, 343–358 (2000)

I.C. Mori, Y. Nobukiyo, Y. Nakahara, M. Shibasaka, T. Furuichi, M. Katsuhara, A cyclic nucleotide-gated channel, HvCNGC2-3, is activated by the co-presence of Na+ and K+ and permeable to Na+ and K+ non-selectively. Plants (Basel) **7**, 61 (2018)

Q.A. Ngo, H. Vogler, D.S. Lituiev, A. Nestorova, U. Grossniklaus, A calcium dialog mediated by the FERONIA signal transduction pathway controls plant sperm delivery. Dev. Cell **29**, 491–500 (2014)

T. Niittyla, A.T. Fuglsang, M.G. Palmgren, W.B. Frommer, W.X. Schulze, Temporal analysis of sucrose-induced phosphorylation changes in plasma membrane proteins of Arabidopsis. Mol. Cell. Proteomics **6**, 1711–1726 (2007)

T.S. Nühse, A. Stensballe, O.N. Jensen, S.C. Peck, Large-scale analysis of in vivo phosphorylated membrane proteins by immobilized metal ion affinity chromatography and mass spectrometry. Mol. Cell. Proteomics **2**, 1234–1243 (2003)

T.S. Nühse, A.R. Bottrill, A.M. Jones, S.C. Peck, Quantitative phosphoproteomic analysis of plasma membrane proteins reveals regulatory mechanisms of plant innate immune responses. Plant J. **51**, 931–940 (2007)

Y. Pan, X. Chai, Q. Gao, L. Zhou, S. Zhang, L. Li, S. Luan, Dynamic interactions of plant CNGC subunits and Calmodulins drive oscillatory Ca(2+) channel activities. Dev. Cell **48**, 710–725 (2019).e715

V. Pérez Koldenkova, T. Nagai, Genetically encoded Ca(2+) indicators: Properties and evaluation. Biochim. Biophys. Acta **1833**, 1787–1797 (2013)

S.K. Sanyal, S. Mahiwal, G.K. Pandey, Calcium signaling: A communication network that regulates cellular processes, in *Sensory Biology of Plants*, ed. by S. Sopory, (Springer, Singapore, 2019), pp. 279–309

A.M. Schlegel, E.S. Haswell, Analyzing plant mechanosensitive ion channels expressed in giant E. coli spheroplasts by single-channel patch-clamp electrophysiology, in *Methods in Cell Biology*, (Academic, New York, 2020)

J. Schroeder, R. Hedrich, J. Fernandez, Potassium-selective single channels in guard cell protoplasts of *Vicia faba*. Nature **312**, 361–362 (1984)

S. Shabala, J. Bose, Application of Non-invasive Microelectrode Flux Measurements in Plant Stress Physiology, in *Plant Electrophysiology*, ed. by A. Volkov, (Springer, Berlin/Heidelberg, 2012)

S. Shabala, L. Shabala, I. Newman, Studying Membrane Transport Processes by Non-invasive Microelectrodes: Basic Principles and Methods, in *Plant Electrophysiology*, ed. by A. Volkov, (Springer, Berlin/Heidelberg, 2012)

A. Shankar, A.K. Srivastava, A.K. Yadav, M. Sharma, A. Pandey, V.V. Raut, M.K. Das, P. Suprasanna, G.K. Pandey, Whole genome transcriptome analysis of rice seedling reveals alterations in Ca(2+) ion signaling and homeostasis in response to Ca(2+) deficiency. Cell Calcium **55**, 155–165 (2014)

A. Singh, P. Kanwar, A.K. Yadav, M. Mishra, S.K. Jha, V. Baranwal, A. Pandey, S. Kapoor, A.K. Tyagi, G.K. Pandey, Genome-wide expressional and functional analysis of calcium transport elements during abiotic stress and development in rice. FEBS J. **281**, 894–915 (2014)

H. Sze, F. Liang, I. Hwang, A.C. Curran, J.F. Harper, Diversity and regulation of plant Ca^{2+} pumps: Insights from expression in yeast. Annu. Rev. Plant Physiol. Plant Mol. Biol. **51**, 433–462 (2000)

E. Teardo, L. Carraretto, S. Wagner, E. Formentin, S. Behera, S. De Bortoli, V. Larosa, P. Fuchs, F. Lo Schiavo, A. Raffaello, R. Rizzuto, A. Costa, M. Schwarzlander, I. Szabo, Physiological characterization of a plant mitochondrial calcium Uniporter in vitro and in vivo. Plant Physiol. **173**, 1355–1370 (2017)

M. Thellier, U. Luttge, Plant memory: A tentative model. Plant Biol. **15**, 1–12 (2013)

W. Tian, C. Hou, Z. Ren, C. Wang, F. Zhao, D. Dahlbeck, S. Hu, L. Zhang, Q. Niu, L. Li, B.J. Staskawicz, S. Luan, A calmodulin-gated calcium channel links pathogen patterns to plant immunity. Nature **572**, 131–135 (2019)

M. Toyota, D. Spencer, S. Sawai-Toyota, W. Jiaqi, T. Zhang, A.J. Koo, G.A. Howe, S. Gilroy, Glutamate triggers long-distance, calcium-based plant defense signaling. Science **361**, 1112–1115 (2018)

A. Trewavas, Intelligence, cognition, and language of green plants. Front. Psychol. **7**, 588 (2016)

L.C. van Loon, The intelligent behavior of plants. Trends Plant Sci. **21**, 286–294 (2016)

E.D. Vincill, A.M. Bieck, E.P. Spalding, Ca^{2+} conduction by an amino acid-gated ion channel related to glutamate receptors. Plant Physiol. **159**, 40–46 (2012)

E.D. Vincill, A.E. Clarin, J.N. Molenda, E.P. Spalding, Interacting glutamate receptor-like proteins in phloem regulate lateral root initiation in Arabidopsis. Plant Cell **25**, 1304–1313 (2013)

Y. Wang, Functional Characterization of Plant Ion Channels in Heterologous Expression Systems, in *Plant Electrophysiology*, ed. by A. Volkov, (Springer, Berlin/Heidelberg, 2012)

P.J. White, Characterization of a high-conductance, voltage-dependent cation channel from the plasma membrane of rye roots in planar lipid bilayers. Planta **191**, 541–551 (1993)

P.J. White, Characterization of a voltage-dependent cation-channel from the plasma membrane of rye (*Secale cereale* L.) roots in planar lipid bilayers. Planta **193**, 186–193 (1994)

P.J. White, Calcium channels in the plasma membrane of root cells. Ann. Bot. **81**, 173–183 (1998)

P.J. White, M. Tester, Using planar lipid-bilayers to study plant ion channels. Physiol. Plant. **91**, 770–774 (1994)

A.K. Yadav, A. Shankar, S.K. Jha, P. Kanwar, A. Pandey, G.K. Pandey, A rice tonoplastic calcium exchanger, OsCCX2 mediates Ca^{2+}/cation transport in yeast. Sci. Rep. **5**, 17117 (2015)

E. Zakharian, Recording of ion channel activity in planar lipid bilayer experiments. Methods Mol. Biol. **998**, 109–118 (2013)